Feedback
Control
Systems
for
Technicians

Feedback Control Systems for Technicians

Robert T. Pickett

Piedmont Virginia Community College

PRENTICE HALL, Englewood Cliffs, N.J. 07632

Library of Congress Cataloging-in-Publication Data

Pickett, Robert T.
  Feedback control systems for technicians/Robert T. Pickett.
  p. cm.
  Bibliography: p.
  Includes index.
  ISBN 0-13-313933-6
  1. Feedback control systems. I. Title.
TJ216.P53 1988
629.8'3–dc 19

Editorial/production supervision and
  interior design: Ed Jones
Cover design: 20/20 Services, Inc.
Manufacturing buyer: Ed O'Dougherty
  and Margaret Rizzi

Cover photograph courtesy of
  AEP/Appalachian Power

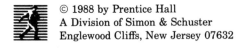
© 1988 by Prentice Hall
A Division of Simon & Schuster
Englewood Cliffs, New Jersey 07632

Printed in the United States of America

10 9 8 7 6 5 4 3 2 1

ISBN   0-13-313933-6   025

Prentice-Hall International (UK) Limited, *London*
Prentice-Hall of Australia Pty. Limited, *Sydney*
Prentice-Hall Canada Inc., *Toronto*
Prentice-Hall Hispanoamericana, S. A., *Mexico*
Prentice-Hall of India Private Limited, *New Delhi*
Prentice-Hall of Japan, Inc., *Tokyo*
Simon & Schuster Asia Pte. Ltd., *Singapore*
Editora Prentice-Hall do Brasil, Ltda., *Rio de Janeiro*

to
my darling Lilli

Contents

Contents

Preface

This book is for anyone who wants to learn what closed-loop control is all about; it is particularly for the technician who is working on control systems for electric power plants, chemical processes, conveyor belts, knitting machines, or any other application of automatic control. The technician is the engineer's "right hand" and needs to understand the fundamental ideas of automatic control, the behavior patterns of closed-loop control systems, some of the problems likely to arise, and some of the possible solutions. The technician is usually responsible for much of the equipment assembly and installation and is involved with testing and troubleshooting. Thus it is vital to have a good understanding of control behavior and of practical matters of installation, maintenance, and troubleshooting.

Although not a designer, the technician can be of much help in solving or avoiding problems; after all, day-to-day experience in plant operations is a great learning experience. The technician should endeavor to learn how control systems are designed and components selected, and be ready to assist when control problems arise and when new control equipment is being considered. Particularly valuable are ideas on reliability, safety, maintainability, and ease of troubleshooting. These last topics are rarely covered adequately in engineering schools, and are learned chiefly from experience. The technician will be able to help write specifications that will avoid future problems with inadequate equipment.

The technician therefore needs to learn as much as possible about closed-loop systems, components, and behavior, without necessarily understanding higher-mathematics design techniques. This book endeavors to point out the relationships between overall closed-loop behavior and the characteristics of the control components, and also some practical matters of installation and troubleshooting. The treatment is essentially nonmathematical, with graphs and descriptions rather than equations.

The first chapters are concerned with the ideas behind closed-loop control; later chapters deal with practical hardware problems such as adequate specifications and proper installation. Next, as examples of very large scale complex controls, the control requirements for large utility electric plants are outlined—both conventional (coal/oil/gas fired) and nuclear power plants. Finally, some of the latest control techniques and devices are described: programmable controllers, "smart" transmitters, fiber optics, and network operations. Three vital topics are discussed in the appendices: operational amplifiers, optical isolators, and cable shielding.

This is not a design book; the intent is to explain the behavior of closed-loop systems, outline potential problem areas, and suggest many practical matters that must be considered in order to obtain satisfactory, reliable closed-loop control. If detailed design information is required, the reader is advised to consult specialized texts, some of which are listed in the Bibliography.

This material has been used in one of the last courses of our two-year Associate in Applied Science Degree program. As prerequisites, all students must have completed courses in ac/dc analysis, electrical measurements, solid-state circuit design, digital circuits, motors, transformers, and two quarters of microprocessor programming and interfacing using assembly language.

Many thanks are due to the following people, who were so cooperative and helpful in supplying photographs and product information: Jack Magar, Larry Worth, and Ken Wilt of Honeywell (Industrial Programmable Control Division), Joe Roantree of Honeywell (Industrial Controls Division), Anne Grier of Virginia Power, and Dick Burton and Charlotte Lavinder of Appalachian Power.

Robert T. Pickett

Feedback
Control
Systems
for
Technicians

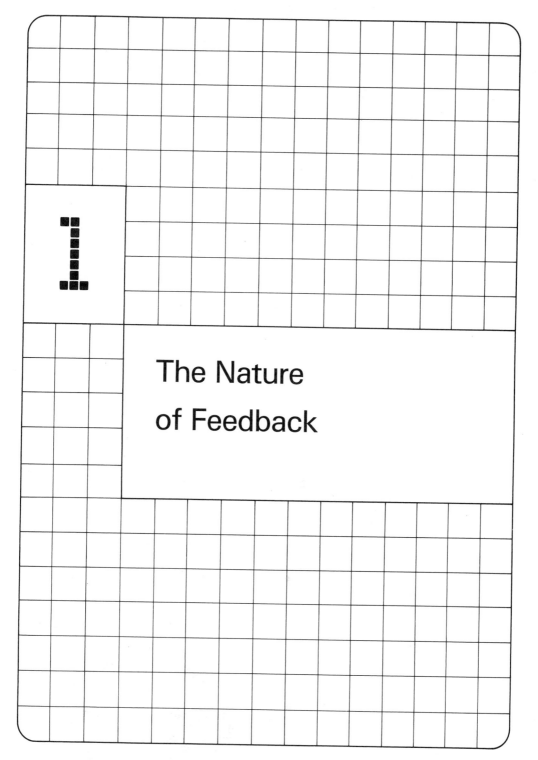

1

The Nature
of Feedback

OBJECTIVES

To explain the differences between open- and closed-loop control systems

To show a diagram of a typical closed-loop system

To explain the advantages of closed-loop control

To discuss several examples of open- and closed-loop systems

To indicate some difficulties and disadvantages of closed-loop controls

CONTROL SYSTEMS IN GENERAL

This book is about feedback control systems: what they are, how they behave, and how to install and keep them operating properly. It is particularly for the person who does the installation, maintenance, and troubleshooting when something goes wrong. Most machines and processes today are controlled using some type of "feedback," which is the topic of this chapter.

Control of some sort is required for everything. In some cases, inherent control is present—that is, nothing extra (man-made) is required to limit the behavior of a machine. An example of self-control is the boiling of water; the temperature of the water rises to 100°C (at sea level) as heat is applied, and then the temperature remains constant at 100°C. Natural behavior results in an equilibrium between vapor pressure and atmospheric pressure, which limits the temperature to 100°C.

Other systems show little or no tendency toward self-control. An example is an oil-fired home heating furnace (see Fig. 1-1); such a furnace, without some form of burner control, is likely to overheat both itself and the house and possibly cause damage. Nothing is present to

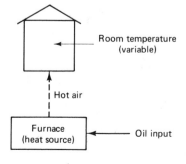

Fig. 1-1 Home heating system, open-loop.

regulate the fuel input according to house temperature; such a system is likely to be a nuisance and may even be dangerous. This is an example of an open-loop system: The burner creates an input (heat source) and the output is hot air, which results in room-temperature changes. Nothing exists to turn the burner on and off according to room temperature, it is "one-way," hence *open loop*, because there is no feedback "loop" to compare what is happening with what should be happening.

When a room thermostat is added (see Fig. 1-2), a *closed loop* is created whereby the burner is turned on or off according to room temperature. Note that the oil input to the furnace depends on a comparison of actual room temperature with desired temperature; the comparison gives an error signal. The controller then opens the oil valve if the error signal indicates that the room temperature is lower than desired. This is a feedback system, in which an error signal controls system behavior. It is called "closed loop" because the output signal is looped back to compare to the desired signal.

OPEN-LOOP SYSTEM BEHAVIOR

Another open-loop system, similar to the uncontrolled furnace example above, is a gasoline engine coupled to a variable load (see Fig. 1-3). The engine will change speed as the load is changed. If a constant speed is required, constant operator attention is necessary. The throttle position depends on the speed desired and the load imposed. If the load is absolutely constant, the throttle position can be calibrated in terms of speed, but even then the engine speed is not likely to remain constant for long.

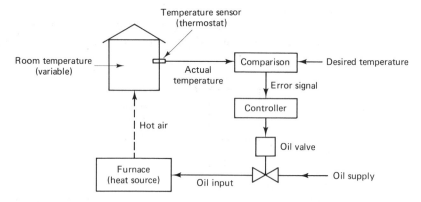

Fig. 1-2 Home heating system, closed-loop.

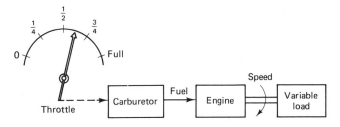

Fig. 1-3 Gasoline engine: open-loop operation.

The difficulties of such open-loop speed control are as follows:

1. Periodic calibration is required to indicate that each throttle position causes a known speed. Figure 1-4 shows a calibration curve for speed versus throttle position. Note that the two curves represent full and light loads. These would be called *linear* calibration curves (meaning "straight"). More realistic calibration curves are shown in Fig. 1-5; these curves are not straight and are called *nonlinear*. For example, one-fourth throttle may give 1000 rpm, but one-half throttle may give 3600 rpm and three-fourths throttle may give 4800 rpm.

2. Many parameters, such as engine friction, will change as the engine warms up. Also, the quality of gasoline (octane) may change and affect the speed.

3. Load variations change everything, so that calibration has to be repeated. In other words, it is practically impossible to predict what engine speed will result for a given throttle setting.

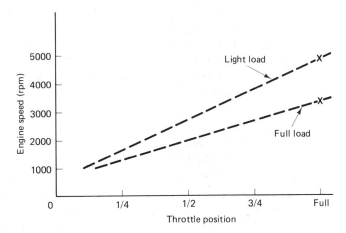

Fig. 1-4 Gasoline engine: linear speed versus throttle position.

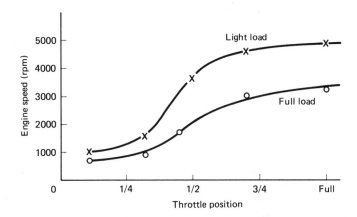

Fig. 1-5   Gasoline engine: nonlinear speed versus throttle position.

One way to avoid these shortcomings is to attach a tachometer for an operator to observe, and then a closed-loop system results, with the operator moving the throttle to correct for speed changes. The operator "closes the loop," moving the throttle according to the speed error.

CLOSED-LOOP SYSTEM BEHAVIOR

Continuing with the example used above, the gasoline engine speed control: If we add the tachometer to measure the output speed and couple that measurement to the throttle, we have an automatic closed-loop speed regulating system (see Fig. 1-6). The throttle is automatically adjusted to maintain the desired speed. The tachometer reading is sent to an error detector which compares actual speed to desired speed, and a controller moves the throttle as required to maintain the desired speed. Note the closed loop; the throttle is moved in response to an error signal.

The closed-loop system has the following advantages:

1. Speed can be kept constant despite load variations.
2. Component nonlinearities are less troublesome; closed-loop control will move the throttle until the proper speed is maintained, even if more movement is required than expected.
3. Accuracy is maintained; of course, this depends on the accuracy of the tachometer. Engine speed can now be depended on to remain nearly constant despite variations in load, friction, efficiency, gasoline octane, and so on.

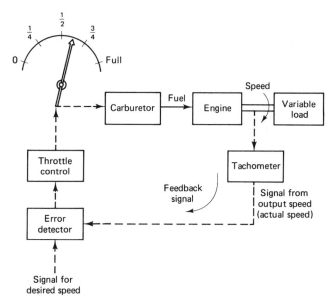

Fig. 1-6 Closed-loop operation of a gasoline engine.

TYPICAL OPEN-LOOP SYSTEMS

Traffic lights (the old-fashioned variety) are a notorious example of open-loop systems; nothing affects the time sequence, and traffic movement has nothing to do with the planned sequence of red, amber, and green. The lights themselves run smoothly whether or not the traffic does. Worse still, many cities have traffic lights operating at road intersections with no coordination to the next intersection. That is, each intersection is independent; traffic could leave one intersection and be stopped needlessly at the next. Some older systems have attempted to coordinate the light timing so that, theoretically, if a vehicle moved at 28 mph, for example, all lights would turn green as approached and traffic could move continuously on a main route. It is obvious, however, that these traffic lights are properly timed for only one average vehicle speed, so that when traffic gets heavy, the system of traffic lights may actually aggravate the problems rather than improve traffic flow, as intended.

These older systems consist of a motor-driven cam arrangement in which each cam opens and closes switches as the assembly rotates (see Fig. 1-7). Timing is changed by installing different cams for different on/off cycles, or changing the speed of the drive motor. Any desired change in the length of red/amber/green cycles requires a person to travel to the intersection and adjust the timing. Trial and error is involved, and someone must observe traffic for a few days to determine if timing changes

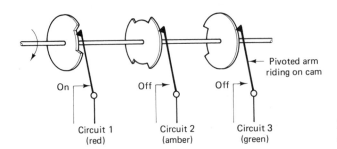

Fig. 1-7 Mechanical timer with cam-operated switches.

made traffic flow better or worse. Note that this traffic light control system is an automatic control but that no feedback loop is used. It is thus an open-loop system.

Fortunately, traffic systems are now being installed that measure traffic flow and time the lights accordingly. Central control is being developed (using computers) which coordinates traffic flow, intersection to intersection. This is a move in the right direction. The desired result is smooth traffic flow—not necessarily smooth operation of traffic lights.

TYPICAL CLOSED-LOOP SYSTEMS

Inherent Feedback Loops

Closed loops, in one form or another, are usually present in natural systems; water boiling at 100°C is an example. In biological systems, animals grow to a fixed size depending on their species. Nature has many inherent controls on size, quantity, speed, and even population growth. However, the natural regulating tendencies are often inadequate for our purposes. Particularly if a machine involves a lot of power, external control may be mandatory. Nature might eventually regulate, but perhaps not before something exploded or flew apart from high speed. For example, in a gasoline engine with a wide-open throttle, friction might eventually limit the speed, but the engine might fly apart before reaching that limit. Nature might regulate the speed if the materials could stand the speed.

External Feedback Loops

Several examples of external loops follow.

   1. *Speed control using a flyball speed governor* (see Fig. 1-8). This scheme was one of the first speed control systems; here we have a simple throttle-regulating device, moving the throttle to accomplish constant-speed operation. The metal balls (a) are hung on light metal rods attached to the shaft at (b) and to a sliding metal collar at (c). As the balls

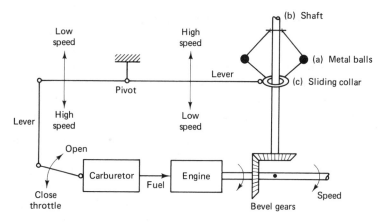

Fig. 1-8 Flyball speed governor.

rotate they tend to fly out and thus raise the sliding collar, which is connected to the throttle through levers. As the balls rise (increasing speed) the throttle is closed, and as the balls drop, the throttle is opened.

2. *House temperature control using a simple feedback loop with an on/off thermostat* (see Fig. 1-2). The furnace is simply turned on or off according to the temperature error. This type of control always allows the house temperature to vary perhaps 5 degrees or more because of the turn-on, turn-off differences and the time lags involved in heating up the furnace and house. Nevertheless, it is usually considered an adequate scheme for the purpose.

3. *Power steering of automobiles.* This is a scheme that utilizes external power to aid in turning the wheels. In this case, power is applied to move the wheels proportionately to the amount of steering wheel movement. Various variations are used: full power, aided manual, and so on. "Full power" means that all power for turning the wheels comes from the power steering motor or hydraulic system. The driver does not work at all, and the steering wheel is very easy to turn. Partial, or aided-manual steering requires the driver to supply perhaps 30% of the effort needed to turn the wheels, with the remainder supplied by the power system. In this scheme the resistance to turning can be felt by the driver. If, for example, the wheels begin to slip on ice, there is a difference in the steering wheel's resistance to turning. This is a safer system than full power because the driver can feel road conditions before the vehicle is out of control. "Road feel" is a term understood by most people but very difficult to measure scientifically. People may simply prefer one automobile to another because of road feel.

4. *Cruise control on automobiles.* This is a familiar form of closed-loop system; the throttle is automatically adjusted to maintain a desired speed. A speed sensor (on the speedometer cable or transmission) generates a signal proportional to actual speed; an electronic error detector compares the actual speed to the desired speed, and a controller sends a demand signal to the actuator (usually a pneumatic device operating on engine vacuum) which adjusts fuel flow to the engine.

5. *Machine tool positioning.* The so-called "automation age" depends on feedback loops. In particular, machine tool position commands may be coded on magnetic tape. Feedback is used to check to see if a machine actually reaches each position commanded. If not, corrections are made automatically. This principle of comparing commanded with actual positions is the vital point that forms the basis of automatic control.

6. *Gun platform on a ship.* Automatic control is required to correct the platform position when ship movement, gun firing, and other conditions disturb the aiming of the gun. The objective is to maintain a particular gun direction and elevation despite all disturbances. Closed-loop systems with fast response and high horsepower are needed for this application.

COMPONENTS OF A CLOSED-LOOP SYSTEM

The essential components of a closed-loop control system are shown in Fig. 1-9:

1. A process, which is what we are controlling. It may be an engine, a house furnace, a gun platform, a water pressure system, a chemical plant, or a nuclear power plant.
2. Measuring devices (sensors), which measure the controlled variable and send a signal to the error detector. (These devices sense the pressure, temperature, flow, and speed of whatever variable is to be controlled.)
3. An error detector, which receives the measured signal and compares it with the desired (set-point) signal. The difference is the error signal.
4. A controller, which decides what to do with the error signal and sends out signals to an actuator, such as a hydraulic cylinder.
5. A control element (actuator), which is the final link and accomplishes what the controller calls for. It may be an electric motor, a hydraulic cylinder, or a pneumatic device.

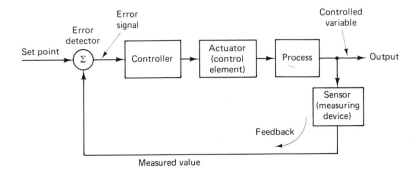

Fig. 1-9 Essential components of a closed-loop control system.

A complete loop with feedback is thus present: The output is compared to the input, generating an error signal that is used to correct the output.

An important point worth emphasizing is that every automatic control system is not necessarily a closed-loop system; the traffic light system discussed earlier had *no* feedback from actual traffic flow to adjust the timing of the lights and thus is an open-loop system. An error detector must be present to create a closed loop.

Another very important point is that closed-loop systems are systems, not merely collections of components picked from catalogs. Proper behavior depends on coordinating the behavior of process, sensors, controller, and actuator. If they are not coordinated, overall behavior may be worse than desired or even totally unsatisfactory. For example, if the engine speed control system is set to move the throttle full-open for a very small speed error, the engine will speed up very rapidly, causing the controller to respond by closing the throttle to reduce speed. Next the controller would open the throttle because the speed is low. The result is a continuous speed-up and slow-down cycle (oscillation), which would be unacceptable (see Fig. 1-10).

ADVANTAGES OF CLOSED-LOOP SYSTEMS

Closed-loop systems have the following major advantages:

1. They can act faster than any human being can, in places where human beings cannot live—such as areas of high or low temperature, in outer space, or in areas with nuclear radiation.

2. They can do a job faster than human beings can. Automatic systems can make decisions and initiate action in milliseconds, whereas a person might require several seconds to decide what to do. Some

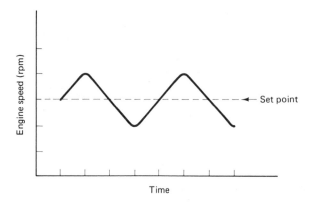

Fig. 1-10 Unsatisfactory behavior of a speed control system.

chemical processes cannot be controlled manually because temperatures and pressures move too rapidly.

3. They are consistent and usually more dependable than people.

4. They can take into account more variables than one person can easily comprehend. Many automatic control systems take into account several temperatures and pressures before making actuator movements. Watching several input signals and then deciding what to do may be almost impossible manually.

5. They may be adjusted to optimum control performance and thereafter depended upon. Automatic controls do not become fatigued and do not change their behavior as may happen with different operators in a control room.

DISADVANTAGES OF CLOSED-LOOP CONTROL

Are there disadvantages of closed-loop control? There certainly are, among which are the following:

1. Equipment is more complex and more expensive than that of simple manual open-loop controls.

2. Installation is more difficult and adjustments are more critical than with manual controls.

3. Equipment maintenance is more difficult because of the electronics involved.

4. Equipment maintenance requires more knowledgeable people, with good training.

5. There is a possibility of runaway due to a faulty error signal; that is, the actuator runs continuously until some mechanical limit is

reached. This can happen if anything in the closed loop causes the error signal to remain high, constantly. There are many types of failures that can cause this: open or shorts in wiring to the sensor, open or shorted wiring to the set-point command input, defective error detector or controller circuitry, or faulty actuator control. Power supply failures can cause this also; for example, if a controller runs on two voltages, $+15$ V and -15 V; if one of these fails, the controller may output a full-speed demand signal to the actuator.

Because of this possibility, a separate safety system should always be installed, so that the process can be shut down automatically before damage occurs due to a runaway.

SUMMARY

Closed-loop systems are widely used because of the many advantages; however, there can be problems which can be minimized with careful design and provision of separate safety systems.

STUDY QUESTIONS

1. How can you determine if a control system is open or closed loop?
2. What are three advantages of closed-loop systems?
3. Are there advantages to open-loop control?
4. Draw a block diagram of a closed-loop system, showing the vital functions.
5. Give an example of nonlinear control behavior.
6. What difficulties can be caused by nonlinear control characteristics?
7. Are there disadvantages in using closed-loop control?
8. Explain how an automatic control may not necessarily be a feedback control system.

2

Typical
Control Systems

OBJECTIVES

To describe typical closed-loop control systems
To outline control requirements for an industrial steam boiler
To discuss requirements for a web handling process
To indicate requirements for safety systems

In this chapter we illustrate typical feedback control systems used in industry. In addition, the purpose and general considerations of safety systems are outlined.

WATER BATH TEMPERATURE CONTROL

A constant temperature is required for many types of chemical processes; one way to accomplish this is to surround the chemical tanks with water and control the water temperature. Processing of photographic film is one example. A water bath temperature control system might be constructed as shown in Fig. 2-1; the tanks of photochemicals (1, 2, 3) are surrounded by water, which is heated by the electric heater in the bottom. The controller compares the actual temperature to the set-point temperature and turns on the electric heaters if the actual temperature is below the set-point temperature. The object is to maintain the water temperature as close as possible to the desired set-point temperature.

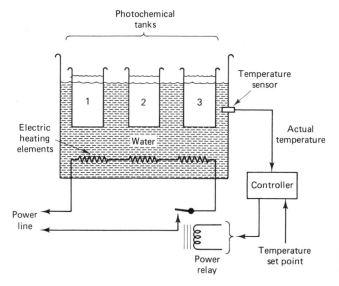

Fig. 2-1 Water bath temperature control.

17

The simplest temperature sensor would be an electrical thermostatic snap switch, which is closed until the water bath temperature rises above the desired set point. The electrical heater would be wired in series with the relay as shown. The water bath temperature behavior would be as shown in Fig. 2-2. Note the continuously changing temperature (called cycling) that results because the heater is either on or off. This continuous swinging of temperature may be acceptable for some situations but not good enough for others, which might require the temperature to be controlled within 1 degree of the set point. With an on/off control system, if the set point is 105 degrees, the actual water temperature might vary over a range of 100 to 110 degrees.

If better temperature control is required, a different type of temperature sensor must be used, such as a thermocouple, which gives a voltage proportional to temperature. That temperature signal goes to an error detector, then to a controller that controls the amount of heat; the amount of heat depends on the amount of error. In this manner a small temperature error would call for a small amount of heat input and a large temperature error would cause a large heat input. This system could provide much better temperature control than the on/off system first described above. The temperature could possibly be maintained within 2 degrees of the 105-degree set point, for example.

STEAM BOILER CONTROLS

Many small boilers are operating to supply steam for heating purposes, as well as for steam turbines that drive electric generators. ("Small" refers to the steam flow output; under 50,000 lb/hr is considered small, compared to the units discussed in Chapter 7, which can generate over

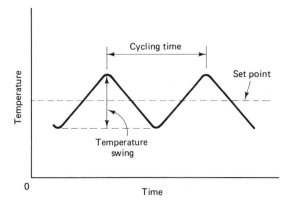

Fig. 2-2 Temperature cycling using on/off control.

8 million pounds per hour). These small boilers provide an independent source of electricity in case of failure of normal electric power from the utility. It is usually impractical (and too expensive) to generate all of the electricity needed for a factory, but it is feasible to generate enough for emergency purposes: some lighting, and water pumps for firefighting. In the last few years, several electric utility systems have had major failures, during which regions of several states lost all electricity. Power was off for several hours—a nuisance to households and a serious problem to some industrial plants. Chemical refineries, for example, must keep material moving in a pipe; also, some chemical reactions generate heat, and a continuous flow of cooling water is mandatory to avoid damage to equipment.

Another possibility is the fact that the normal source of electricity may have failed due to a storm or flood; that is, of course, the time during which water pumps must be operating to prevent damage to machinery. A standby source of electricity is very useful, to say the least.

Some industries install standby diesel generators, but these are expensive in larger sizes, and also harder to justify economically because they run only occasionally during emergencies. In addition, there will be a period of perhaps several minutes before the diesel engine is able to start and reach full operating speed and power. Electric power may come back on with a surge, which may cause problems.

A small steam boiler, on the other hand, can run all the time, supplying steam for heating, and also for running a steam turbine driving an electric generator. The generator could supply some of the power required by the plant and be ready to supply emergency power when necessary (because it is running all the time, electric power would not be lost completely when the utility source fails). There is considerable advantage in having a continuous source of electricity on your own premises.

There is also the possibility that electric power can be sold to the electric utility company and thus help pay for the boiler-turbine-generator installation. This is particularly attractive if a large amount of waste material is available to use as fuel, such as wood bark at a paper mill or waste oil at a refinery. These fuels are essentially free, and must be disposed of somehow. Generation of electric power for all plant needs, and also for sale, can be a real benefit.

Boiler Construction

A typical oil-fired boiler (very simplified) is shown schematically in Figure 2-3. The major variables to control are steam pressure and steam-drum water level. Feedwater comes into the lower drum pumped by the

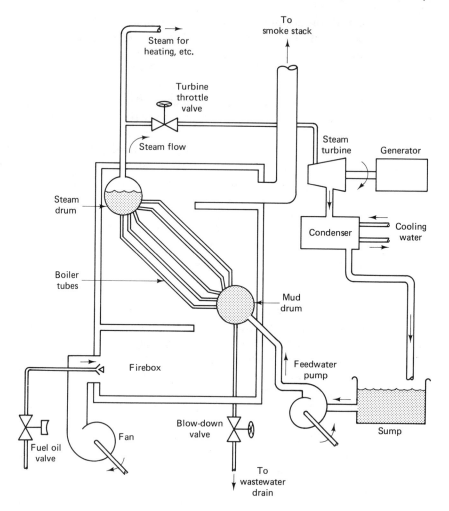

Fig. 2-3 Small steam boiler.

feedwater pump; pump speed is varied to control flow. The turbine throttle valve controls the flow of steam to the turbine and thus the amount of electricity generated (kilowatts). Exhaust steam from the turbine is cooled in the condenser (back to water) and flows to the sump, to be pumped around again by the feedwater pump. Steam is also used for heating and in some cases may be condensed and returned to the sump.

Note that this feedwater must be extremely pure, because impurities (calcium, iron, etc.) could deposit on the turbine blades and in the pipes in the boiler. Such deposits can clog up the boiler pipes and cause major damage due to unbalance in the turbine. The lower drum is generally called a mud drum, because sediment collects there (even with very pure water). A "blow-down" valve is opened regularly to dispose of that sediment. A makeup water system (not shown) is required to supply the ultrapure water lost in blow-down and also the water lost for steam heating if that water is not returned to the sump.

Feedwater pumps may be driven by electric motors or steam turbines. Many boiler systems have two feedwater pumps, one electric and the other steam driven; reliability is thus improved, inasmuch as feedwater flow can be maintained even if electric power is lost. Feedwater flow is very important because severe damage to boiler tubes can occur due to overheating if water flow is not sufficient.

A very important variable to control is boiler drum water level. If the level is too low, pipes in the furnace may overheat, causing leaks. If the water level is too high, water may be carried over into the turbine with the steam and cause damage (solid water droplets would be like little pellets hitting the turbine blades). Drum level is therefore usually used as a vital input to feedwater flow control, to maintain proper drum level. Steam pressure is controlled by the firing rate (oil flow); to raise pressure, oil flow is increased, which raises furnace combustion temperature and thus steam pressure. Proper combustion of the fuel oil depends on an adequate supply of air and is generally a fixed ratio to the combustion airflow. Proper combustion means enough air to burn all of the fuel, leaving no smoke, and not too much air, which reduces the furnace combustion temperature, reducing efficiency.

Other Closed-Loop Controls

Another control is on cooling-water flow, which usually comes from a river or lake. This flow is not critical but must be above some minimum value that will cause complete condensation of the exhaust turbine steam. The objective is to cause a vacuum on the turbine exhaust in order to obtain maximum turbine efficiency, which occurs with maximum pressure drop from turbine inlet to exhaust. The turbine throttle will be controlled by a closed loop, which adjusts the throttle to maintain a desired turbine output and hence the amount of electricity generated. Most of these controls will be automatic closed loops, because it is too difficult, and very boring, for a human being to watch drum level continuously, for example, and adjust feedwater flow.

The most important closed-loop controls are, then:

1. *Feedwater flow:* controlled by steam flow and steam drum level
2. *Steam pressure:* controlled by fuel flow

3. *Combustion airflow:* controlled by fuel flow
4. *Electrical load:* controlled by steam turbine throttle

Feedwater flow is very important and is usually directly controlled by steam flow—in other words, feedwater flow, in pounds per hour, is forced (by a closed loop) to equal steam flow, in pounds per hour. The drum level is then used to adjust feedwater flow (up or down) as required to maintain desired drum level.

Fuel flow may simply be controlled by steam pressure error (that is, if steam pressure is lower than desired, the control raises fuel flow). More sophisticated systems control fuel flow according to steam flow and then adjust fuel flow up or down if the steam pressure is not correct. Of course, this is a better control system because fuel flow is adjusted when steam flow changes, before waiting for steam pressure to vary; the object is to maintain constant steam pressure, even though the steam flow varies.

Combustion airflow is adjusted to maintain the desired ratio of air to fuel, for complete combustion; the ratio depends on the heat content of the fuel used, Btu (British thermal units) per pound. The airflow control may be done with a damper in the air ducts, or by speed changes in the drive motors. The electrical output of the turbine is controlled by the throttle, either manually or with a closed loop that maintains a desired power output from the turbine.

WEB HANDLERS

Many industrial processes handle material in continuous strips, such as paper, cloth, or steel. A sketch of a typical process is shown in Fig. 2-4; material is unwound on the left and rewound on the takeup reel on the right. The float idlers move up or down, depending on the speeds of the drive motors. If the material stretches, one of the idler floats will drop, requiring a speedup of the next motor. If the process is a steel rolling mill, the material becomes thinner and thinner, hence moving faster and faster; the take-up motor, 3, must run faster than motors 1 and 2. This is a very difficult control problem because the three motor speeds must be accurately controlled to maintain the proper idler float positions, over a full operating speed range from startup to maximum material speed, maybe over 2000 ft/min. The position of the float idlers must be measured very accurately and quickly, and motor speeds corrected quickly. Many closed-loop controls are involved, and must be carefully designed and accurately adjusted.

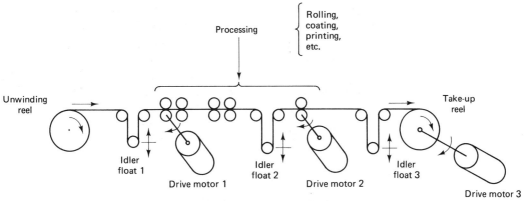

Fig. 2-4 Web handling process.

If the material is paper, as in a high-speed newspaper printing press, breakage is a problem if motor speeds are not accurately controlled. Even more complex are color printing presses, which print three or more colors separately; each color impression must be placed correctly over the previous color impressions to create a true color print. Lining up these impressions requires close control of the idlers and motor speeds, as well as paper position, left and right.

SAFETY SYSTEMS

A safety system design philosophy should be developed and written down, so that all design personnel and technicians understand. Basically, safety system guidelines might be as follows:

1. Safety systems must have separate sensors, so that failure of a normal control sensor will not also disable the safety system.
2. Safety systems must have a separate power source, preferably battery backed, so that plant shutdown can be accomplished if normal control power is lost (despite emergency generators).
3. Safety system wiring should preferably be routed separately from normal control cables to avoid disabling the safety systems if control cables are damaged. For example, never include safety functions in the same cable with normal control signals.
4. Separate actuators must be provided for safety in some cases, depending on the process involved. The designer must always ask what might happen if normal control loops operate incorrectly; how can the safety system cause an orderly shutdown, especially if the

normal actuator fails or if normal line power source fails? Answers to these questions depend on the particular installation.

5. Testing and preventive maintenance of the safety system must be done regularly, because under normal operating conditions, the safety equipment does *nothing* but monitor the operating system. How do you know that it will operate in case of a real emergency?

It is very important to consider safety factors, especially if high-energy processes and large machinery are involved.

Boiler Safety Systems

A steam boiler would have at least the following safety shutoff features:

1. Over-pressure steam
2. Flame failure
3. Low steam-drum water level
4. Over-temperature in stack
5. Low pressure in steam line
6. Low lubrication oil pressure

Each of the items above is based on *safety*. Excessive steam pressure must be avoided, due to the possibility of explosions. Flame failure is very dangerous, because explosions can result if fuel oil strikes the hot surfaces in the firebox. A low water level indicates insufficient feedwater flow and possible damage to the boiler tubes due to over-temperature. The tubes in the furnace are cooled by the presence of water inside, generating steam bubbles that rise up to the steam drum. Boiler tubes can actually melt if not filled with water continuously.

Over-temperature in stack gases can indicate excessive fuel input or insufficient air for combustion. Low-pressure steam can indicate feedwater pump failure or excessive air in the furnace. In any case, low steam pressure may indicate that water droplets may be carried over into the steam turbine, causing damage. Lubrication oil pressure is vital; otherwise, bearings in the turbine-generator can be damaged.

Machinery Safety Systems

Consideration should be given to the following safety monitors for motor- or turbine-driven machinery:

1. Over-speed
2. Reverse-phase rotation

3. Excessive current
4. Over-temperature
5. Vibration

Over-speed protection is necessary when using machinery driven by direct-current (dc) motors or steam turbines. A DC motor can reach dangerous speeds if the field current supply is lost; a steam turbine may exceed a safe speed if the throttle sticks open for any reason or if the load suddenly drops to zero, perhaps due to a broken shaft coupling.

The electrical phase rotation determines the direction of rotation of three-phase motors, and hence the motors will run backwards if phase rotation is reversed. Reverse-phase rotation can occur on the utility power lines to a plant, when those lines are repaired after a storm and wires are replaced somewhat differently from before the storm. It can also occur within a plant when power distribution lines are repaired or rerouted, or transformer banks are reconnected.

Excessive current to an electrical motor can occur because of several conditions: excessive load, due to defective machinery, lack of lubrication, or defective bearings. If the maintenance technician is aware of excessive current, then perhaps the problem can be remedied before a fuse blows, shutting down some machine.

Excessive temperature of motors and other equipment may be an indication of trouble due to overloads or defects. If the technician is aware of this, there may be some way to correct the situation before damage occurs or automatic shutdown occurs.

Vibration and noise in a machine usually indicate trouble due to machinery imbalance. Particularly in large steam turbines this is a vital monitoring function because of the rotor weight (several tons) rotating at 3600 rpm or more. Noise monitors can indicate bearings that are failing, before any major damage occurs.

Safety System Backup

Normal operation is one thing, but what happens if a control loop and the safety system operates improperly? There should be some independent system which monitors important process variables and can initiate a plant shutdown if dangerous conditions occur and the regular safety system does not prevent hazardous operation.

The steam line safety valve is a good example (see Fig. 2-5). It is a separate mechanical device that opens to prevent boiler explosions if steam pressure is excessive. The safety valve is a last-ditch type of device, and really is not expected to be used, because there are other pres-

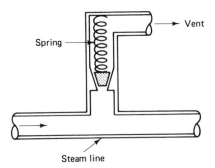

Spring

Vent

Steam line

Fig. 2-5 Steam pressure safety valve.

sure sensors and systems which are *supposed* to prevent excessive steam pressure. Steam overpressure can result in an explosion and is always a serious concern in boiler operation. There is always a safety valve, which operates to reduce pressure if dangerous levels are reached. The spring pressure is adjusted so that the valve opens at the desired pressure. These valves are supposed to snap open and snap shut, rather than open slowly as pressure increases. The valve should open wide to allow maximum steam flow so that dangerous pressure is reduced quickly; the valve should close quickly when pressure drops because gradual closing may result in damage to the valve seat due to the high-velocity steam flow through the small opening as the valve closes; the valve may then leak all the time.

The normal pressure control system is *supposed* to adjust fuel flow for proper steam pressure. The overpressure safety trip-out is *supposed* to shut down the boiler. The mechanical safety valve is *always* there, just in case everything else fails.

SUMMARY

Closed-loop controls can be very simple on/off systems, or complex systems with many sensors and many variables to control. Steam boilers are widely used in industry, for both steam heating and electricity generation. Several closed-loop control systems are required for safe, efficient operation, as well as safety systems to prevent damage if normal controls fail.

STUDY QUESTIONS

1. Outline the advantages and disadvantages of on/off control systems.
2. Describe the operation of a steam boiler.

3. List the most important control loops on a steam boiler.

4. What are the potential advantages of operating a small boiler on your site?

5. List five boiler safety system inputs that could help to avoid major plant damage.

6. Why should safety system inputs (sensors) be separate from normal operating measurements?

7. In a steam boiler, what could be the result of a low water level?

8. How could reverse-phase rotation occur on a power line?

9. How is reverse-phase rotation protection useful?

10. Why is flame failure in a boiler so dangerous?

11. List several industries that could use web handlers.

12. Describe some of the control loops involved in a web handler.

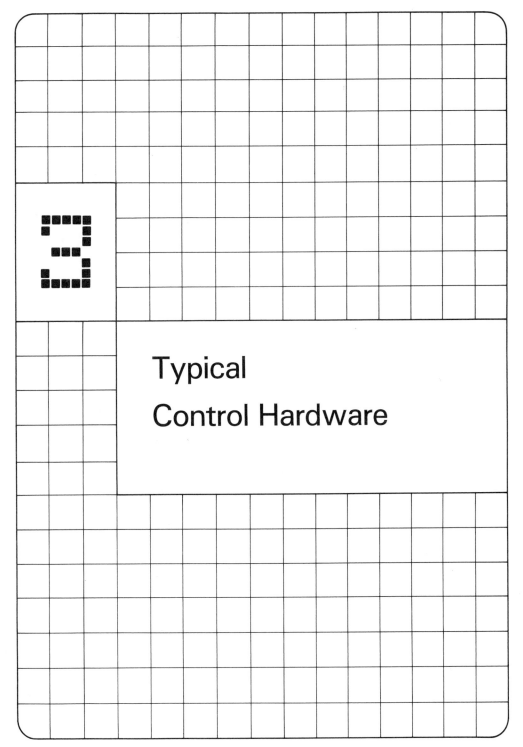

3

Typical
Control Hardware

OBJECTIVES

To describe typical control system hardware: sensors, controllers, and actuators

To explain controller behavior

To compare electric, pneumatic, and hydraulic actuators

To emphasize the importance of proper wiring and cabling

The actual pieces of a control system are called *HARDWARE:* the sensors, controllers, and actuators. Wiring, cabling, and pneumatic signal piping must also be considered hardware, because wiring can introduce noise and pneumatic lines can cause signal delays in a system. In this chapter we describe some of the typical hardware used in industrial control systems.

SENSORS

Sensors provide the feedback signal to a control system; the process variables (pressure, temperature, flow, level, weight, density, viscosity, speed, humidity, distance, vibration, radiation, to name a few) must be measured and converted into some type of signal for transmission to the error detector. Sometimes these are called *transducers,* which is another word for devices that convert one type of variable into another: for example, a fluid flow rate into a digital signal, or a steam pressure signal into a voltage. Pneumatic transducers and control systems use air pressure for signal transmission, typically 3 to 15 psi; measurements and actuator signals are transmitted using air pressure signals.

Electronic sensors are typically designed to use some electrical parameter to indicate the behavior of a physical property. Sensors can be built (some not commercially practical) using any conceivable variable. Some typical operating principles are indicated below.

Parameter	Example
V	Voltage changes can be produced by varying impedance in a circuit or by varying the speed of the generator. In some instances voltage changes are due to other effects, discussed below.
R	Resistance changes can be obtained by changes in temperature or by deformation of the resistive material due to a force.
L	Inductance changes can be accomplished by variation of the magnetic linkages due to displacement.
C	Capacitance changes can be achieved by changes of position between plates or by changes in dielectric effects.

Parameter	Example
F	Frequency changes, alternating current, or pulses can be proportional to speed of a rotating shaft.
Count	A pulse rate can be generated, according to revolutions of a shaft.
Phase	A phase angle can be varied according to movement.

Voltage can be generated in several ways, so many devices (active transducers) have been invented using these basic principles:

Chemical

Electromechanical

Electrostatic

Photoelectric

Piezoelectric

Thermal

Electrochemical devices operate by separating charges. A familiar example is the wet cell constructed with copper and zinc electrodes in a sulfuric acid solution. The zinc electrode collects a surplus of electrons and is thus negatively charged. The cell is a "chemical pump" that pushes electrons to the zinc electrode.

Electromechanical devices, probably the most familiar, generate a voltage by the motion of a wire in a magnetic field. Generators and alternators use this principle, as do tachometers, which are precision generators: Voltage output is precisely proportional to rotational speed.

Electrostatic devices separate charges by friction, as in the familiar physics demonstration in which a glass rod is rubbed with fur, generating a charge on the glass.

Photoelectric devices may be considered in three categories: voltaic, emissive, and resistive. The voltaic device generates a voltage according to the amount of radiant energy received; solar cells are an example. An emissive device emits electrons according to the light received; photodiodes and phototransistors are in this category. The resistive device changes its resistance according to the light received; cadmium sulfide and cadmium selenide are typical materials used to fabricate these devices.

Piezoelectric generation of a voltage occurs when some crystalline materials are stressed. Quartz is a typical natural material that will generate a voltage when distorted by physical pressure. The most familiar application of this principle is in phonograph pickups; motion of the needle (from the record) causes pressure variations on a crystal, resulting in a varying voltage that represents the recorded sound.

Thermal effects can generate voltage; if two different metals are fused together and then heated, a small voltage will be generated (the Seebeck effect). This is the basis of thermocouples, which are widely used in industrial temperature measurements.

Table 3-1 indicates many physical variables that can be sensed and converted into electrical variables.

TABLE 3-1 PHYSICAL MEASUREMENT PARAMETERS FOR TRANSDUCERS

Measurement parameter	Transduction principle
Linear distance (displacement) (dimension) (position)	Capacitance: incremental or differential Inductive: eddy current, variable reluctance Strain gauge: bonded or semiconductor Linear voltage differential transformer (LVDT) Optical: photo electric or laser interferometer Linear potentiometer Linear (digital) encoders

Linear velocity (speed)	Inertial mass: magnetic field, self-generating
	Pendulous mass—spring
	  LVDT
	  Potentiometer
	Inductive
	  Attached linkage
	  Noncontact
	Optical: time differential
	Piezoelectric: integrated acceleration

Linear acceleration	Seismic mass
	  Piezoelectric
	  Piezoresistive
	  Strain gauge
	  LVDT
	  Inductive
	  Capacitance
	  Potentiometer
	Force balance servo

Angular displacement	Capacitor Inductive Resistor Photoelectric Strain gauge LVDT (rotary) Gyroscope Shaft encoder: digital
Inclinometer	Pendulum: potentiometer Force balance: accelerometer, integrator Liquid: resistor

TABLE 3-1 *(cont.)*

Measurement parameter	Transduction principle

Angular velocity (tachometer)	Generator
	  Dc
	  Ac
	  Drag cup
	  Photoelectric or magnetic pulse wheel

Angular acceleration	Electromagnetic + differentiator Force balance servo Digital shaft encoder

Force
  $F = ma$
	Counterbalance
	  Mass
	  Electromagnetic
	Deflection
	  Strain gauge
	  LVDT
	  Piezoresistive
	  Capacitive
	  Inductive
	  Piezoelectric

Torque	Torsional windup
	  Strain gauge
	  LVDT (torsional variable DT)
	  Photoelectric encoder
	  Permeability change
	Dynamometer

Vibration—displacement
  (distance)
  (amplitude)
	Linear displacement transducers (Dc-LVDT)
	Integrated linear velocity transducer signals
	Double-integrated accelerometer signals

Vibration
  Velocity
  Acceleration
  Torsional
	Linear velocity transducers, seismic
	Integrated accelerometer signals
	Linear accelerometers: piezoelectric
	Time interval variations during rotation of each revolution
	Strain gauge: inertial mass

Sound (microphone)	Piezoelectric Capacitance Variable reluctance

Pressure	Bellows: potentiometer
	Capsule differential transformer (LVDT)
	Diaphragm
	  Strain gauge
	  Piezoresistive
	  Piezoelectric
	  Variable capacitance
	  Variable inductance

TABLE 3-1 (cont.)

Measurement parameter	Transduction principle

Flow meters	Positive displacement: volumetric
	  Liquid
	  Gas
	Differential pressure
	
	$$Q = K \sqrt{\dfrac{P_2}{P_1}}$$
	
	  Orifice
	  Venturi
	  Pitot tube
	Turbine: velocity
	  Liquid
	  Gas
	Magnetic: velocity
	Variable area
	  Float meter
	  Force meter
	Thermal: mass flow
	Differential pressure: mass flow
	Turbine—axial: momentum, mass flow

Temperature	Thermocouple
	  K: chromel–alumel
	  J: iron–constantan
	  B: platinum–rhodium
	  T: copper–constantan
	RTD
	  Platinum
	  Nickel
	  Thermistor
	Semiconductor junction
	Pyrometer
	  Radiation
	  Optical

Viscosity
  absolute (poise)
  dynamic (stoke)
	Falling
	  Ball
	  Piston
	Capillary or orifice (Saybolt)
	Rotating member

Humidity	Animal hair: mechanical linkage Lithium chloride: electrical resistance Capacitance Microwave Thermoelectric: optical servo

Source: Table of Physical Measurements appears by permission of Tektronix, Beaverton, Oregon.

PRACTICAL CONSIDERATIONS

Once a transducer is built, it must be calibrated, so that its output signal indicates a known process range; for example, 0 to 10 V might indicate a steam pressure of 0 to 500 psi. The choice of a particular type of transducer depends on the accuracy required, the speed of response desired and degree of ruggedness required, as well as price.

The method of mounting the sensor in the system is an important consideration. For example, a measurement of steam temperature requires cutting a hole in a pipe and welding in a thermometer "well" (see Fig. 3-1). The well must withstand the same pressure as the main pipe, hence must be of high-strength material. If the steam pressure is 2000 psi, the well may be $\frac{1}{4}$ in. or more thick. Thick walls cause a problem, because the temperature sensor will respond slowly rather than following fast changes in steam temperature. This slow-response problem is discussed more fully in later chapters.

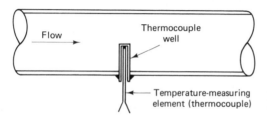

Temperature-measuring element (thermocouple)

Fig. 3-1 Temperature measurement with thermocouple in well.

Other methods have been used, such as mounting the sensor directly inside the pipe, without a well. This method gives faster measurement, but the sensor is likely to be delicate and the fluid flow tends to break the sensor, due to vibrations. Reliability is therefore not good.

Other considerations are accuracy, drift, time lag, delicateness, ease of maintenance, and sensitivity to heat, humidity, dust and dirt, and electrical noise. These topics are all discussed later.

TYPICAL SENSORS

Temperature

Resistance thermometers. *Resistance thermometers* operate on the principle that resistance of most materials varies with temperature. Metals have a positive temperature coefficient; that is, the resistance increases when the temperature increases. Platinum wire is frequently used, wound into a small coil of about 100 Ω resistance; the wire assembly is then enclosed in a protective sheath, as indicated in Fig. 3-2.

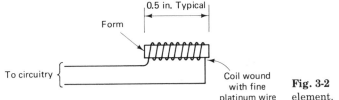

Form

0.5 in. Typical

To circuitry

Coil wound
with fine
platinum wire

Fig. 3-2   Resistance thermometer
element.

Note that the sheath should be as thin as possible, to minimize time delays in heat reaching the sensing element inside. Heavy mounting wells may have a time constant of over 60 seconds, which would not be tolerable in control systems requiring fast response. However, in high-pressure steam systems, heavy wells may be mandatory to withstand the pressure. Figure 3-3 shows a typical well, for mounting a resistance thermometer in a high-pressure pipe. The tip is shown cut away, with the element inside showing. Wiring connections are made in the junction box on the left. Figure 3-4 shows a well of thinner material, for use in lower-pressure systems. This well would have much faster response (shorter time constant) than the one shown in Fig. 3-3.

To convert the resistance changes to an electrical voltage, Wheatstone bridges are frequently used, as shown in Fig. 3-5; when the resistance thermometer element changes resistance due to temperature changes, the bridge will be unbalanced and the voltage across the bridge will indicate the temperature. Simpler circuits are possible, as shown in Fig. 3-6. This simple circuit may actually be superior because it creates

Fig. 3-3   Thermometer well with thick walls. (Courtesy of Honeywell Industrial Controls Division.)

Fig. 3-4   Thin-walled thermometer well. (Courtesy of Honeywell Industrial Controls Division.)

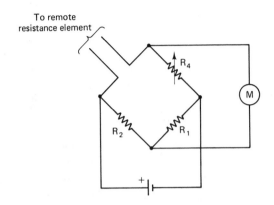

Fig. 3-5 Temperature-measuring bridge.

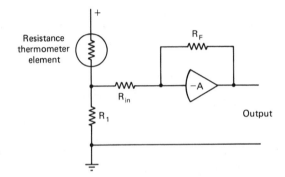

Fig. 3-6 Resistance thermometer with op-amp circuit.

a linear voltage for resistance changes, and operational amplifiers are very stable, dependable devices.

Resistance thermometers using platinum wire have very good linearity, meaning that resistance varies with temperature in a straight line, strictly proportional to temperature. They are rugged, reliable, and can be fabricated to cover a wide temperature range.

Thermistor elements. *Thermistor elements* fabricated from oxides of manganese, cobalt, and nickel, usually have large negative temperature coefficients. Large resistance changes result from relatively small temperature changes, allowing very simple circuits to be used. However, the resistance change is not usually linear, as indicated in Fig. 3-7; however, special compensating circuits can be designed to produce a signal that is almost linear.

Thermistors can be made extremely small, even down to hypodermic needle size. Speed of response is very fast for these small sizes. They can be used bare, or enclosed in glass beads, or in metal sheaths, similar to resistance thermometer elements. They are somewhat delicate and

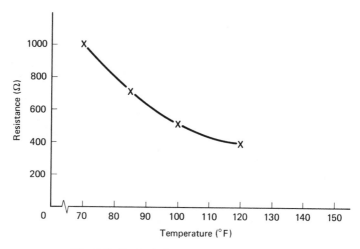

Fig. 3-7 Typical thermistor resistance curve.

easily damaged by excessive current. They are relatively inexpensive and can produce signals of good accuracy with simple circuits. However, care must be taken to avoid self-heating; too much current through the thermistor can raise its temperature and introduce errors.

Thermocouples. *Thermocouples* have been used as temperature sensors probably longer than any other device. They have an almost linear voltage versus temperature characteristic and can be fabricated into rugged packages, suitable for industrial environments. The junctions are usually welded, as indicated in Fig. 3-8. Fastest response is obtained with the exposed junction, but this is fairly delicate, so it is usually enclosed in a sheath or well. The junctions can be mechanically mounted against the inside of the well, for fastest response, but this may lead to electrical ground loop problems and noise pickup, so the insulated version is frequently used (see Fig. 3-9).

Many types of thermocouples are available, fabricated from various combinations of metals. Each has its advantages, depending on the required temperature range and the atmosphere in which it will work. Modern circuits generally involve operational amplifiers, because these

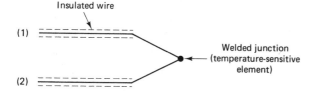

Fig. 3-8 Thermocouple construction. Type E thermocouple: wire 1 is chromel, which is an alloy of nickel and chromium; wire 2 is constantan, which is an alloy of copper and nickel.

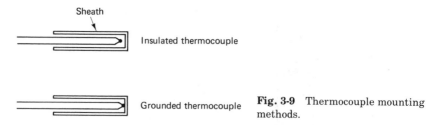

Insulated thermocouple

Grounded thermocouple

Fig. 3-9 Thermocouple mounting methods.

are the best way to obtain usable signal voltages from the thermocouple millivolt output.

Flow

Flow measurements are often made using the venturi principle, in which the liquid flows through a constricted pipe (see Fig. 3-10). The higher velocity through the narrow opening produces a lower pressure; flow can be related to the difference in pressure.

A simpler device is an orifice plate, as shown in Fig. 3-11. Flow is related to the difference in pressure upstream and downstream. The resulting pressure drop can be converted to an electrical signal by any of several techniques. A typical method is shown in Fig. 3-12, in which the diaphragm moves a magnetic element between two coils, which are coupled to the third coil by way of the moving element. This device is called a linear variable differential transformer (LVDT); the unbalanced pickup of the two coils can be converted to a voltage using a bridge circuit. By means of an equation involving a square root, these devices produce a pressure difference signal related to flow that is not difficult to handle, using operational amplifiers.

Another type of flow meter uses a turbine in the pipe, spinning at a speed proportional to velocity of the fluid, which can be related to the mass flow (see Fig. 3-13). The turbine speed is sensed by a magnetic pickup, and electronic circuits count the pulses and compute a signal

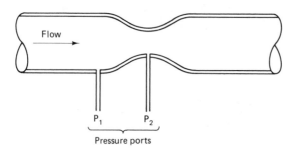

Flow

P_1 P_2

Pressure ports

Fig. 3-10 Flow measurement using a venturi.

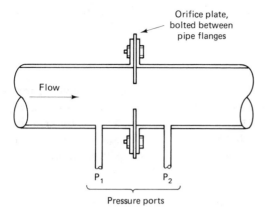

Fig. 3-11 Flow measurement using an orifice plate.

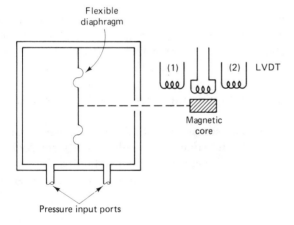

Fig. 3-12 Pressure difference transducer.

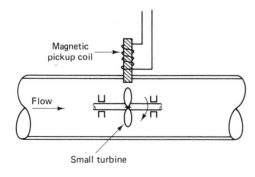

Fig. 3-13 Flow meter using a turbine.

proportional to flow rate. This type of flow meter is widely used in gas flow measurements.

Pressure

Pressure measurement may be accomplished by several methods; one of the simplest is the Bourdon tube principle, indicated in Fig. 3-14. The Bourdon element is a flat hollow tube which tends to straighten out when pressure is applied, moving a dial pointer and the arm on the resistance element. Thus the output voltage is proportional to movement of the Bourdon element.

Other types use a diaphragm element, which moves a magnetic core inside a coil, changing the frequency of an oscillator. The output is then frequency proportional to pressure.

Liquid Level

Liquid level can sometimes be sensed by a simple device such as a float connected to a potentiometer as shown in Fig. 3-15. A clever system for measuring level is the "bubbler," shown in Fig. 3-16. The pressure required to blow bubbles out of the submerged pipe is related directly to liquid depth, and sensed by a pressure sensor. Note that the pressure sensor does not contact the liquid, and thus this scheme is useful if the liquid is corrosive.

Liquid level may also be sensed by conductivity measurements and serve a purpose when the liquid concerned is electrically conductive and nonexplosive. This scheme in its simplest form indicates the level at its set point but does not give a continuous indication of level. A pair of

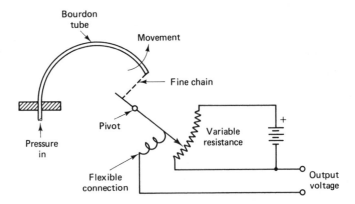

Fig. 3-14   Bourdon gauge principle.

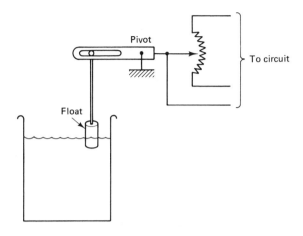

Fig. 3-15   Liquid-level sensor using a float.

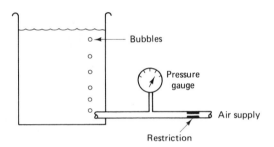

Fig. 3-16   Liquid-level measurement using a bubbler.

electrodes is placed at the desired liquid level so that when the liquid reaches the electrodes, an electrical current flows (see Fig. 3-17).

Radiation techniques have been applied and many variations are possible. One possibility is to place the radioactive source inside the tank at the desired level (see Fig. 3-18). A detector outside the tank is then used to determine when the source is submerged.

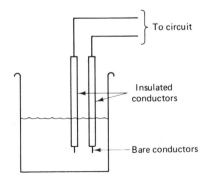

Fig. 3-17   Liquid-level sensor.

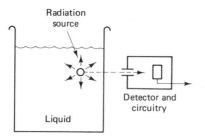

Radiation
source

Detector and
circuitry

Liquid

Fig. 3-18 Liquid-level sensor using a radioactive source.

Speed

A tachometer is the most obvious speed-measuring device, being a precision generator that generates a voltage proportional to rotational speed. When a speed measurement must be added to existing machinery and a tachometer cannot be attached, a magnetic pickup can sometimes be added (see Fig. 3-19). The coil will pick up pulses when the gear teeth pass; electronic circuits can count the pulses and indicate the rotational speed.

If no convenient gear is available for the magnetic sensor, a photoelectric sensor may be used as indicated in Fig. 3-20; a reflective strip is attached to the shaft so that light is reflected into the photosensor once per shaft revolution. Circuits then convert the pulses into an rpm reading.

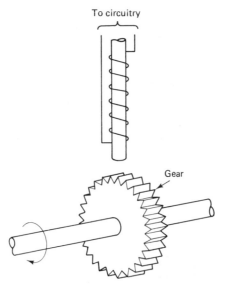

To circuitry

Gear

Fig. 3-19 Magnetic shaft-speed sensor.

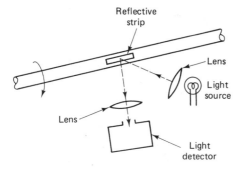

Fig. 3-20 Photoelectric shaft-speed sensor.

Electric Power

Current transformers. An interesting type of transducer is frequently used for electric power monitoring: an ac current-to-ac current transducer, called a current transformer (CT). These devices allow measurement of high-amperage ac current with a lower-range ac ammeter. A typical CT would allow a 5-A meter to read 500 A in the main wire (500:5 ratio). Physically, a CT with this ratio is about 3 in. in diameter.

Typical construction and connections are shown in Fig. 3-21. Note that there is no physical connection between the main conductor and the metering circuit. Thus the CT is useful for reading the current in high-voltage lines.

Construction. A current transformer is built with a single winding on a toroidal (doughnut-shaped) core through which the main current-carrying conductor is run. Many turns of wire are wound on the core. It would seem that this arrangement would produce a voltage stepup, but the secondary is connected to an ammeter, which is essentially a short circuit. The CT is designed to operate with a short-circuited secondary; design of these units is proprietary, and little information has been published regarding design parameters.

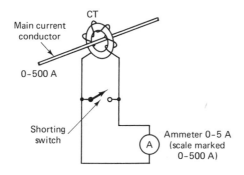

Fig. 3-21 Current-transformer connections.

Precaution: If it is necessary to disconnect the ammeter from a CT, the secondary must be shorted before removing the ammeter. The reason is that the CT has many turns of wire and may generate a dangerously high voltage if operated open circuit. The shorting switch must always be provided when a CT is used.

Potential transformers. Potential transformers (PTs) are precision voltage-step-down units, to allow metering of high-voltage circuits with lower-voltage meters. They are relatively conventional transformers, except for the precise voltage ratios guaranteed (see Fig. 3-22).

Industrial power metering. A typical installation is indicated in Fig. 3-23; volts, amperes, watts, and watthours are metered. The utility system would use the watthour meter (kilowatthours usually!) for billing purposes. Note the shorting switch provided in case the ammeter, wattmeter, or watthour meters must be disconnected.

Summary. Current transformers and potential transformers allow remote metering of high-current, high-voltage circuits. Single-phase circuits are indicated on the figures; three-phase power metering is similar, usually metered with the two-wattmeter method, whereby only two CTs and two PTs are required for three-phase power metering. (See the Bibliography for further information on three-phase power measurements.)

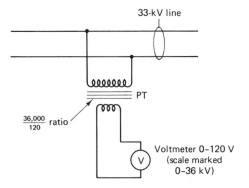

Fig. 3-22 Potential-transformer connections.

Strain Gauges

Strain gauges are a very simple type of sensor, but in practical terms can be very difficult to use correctly. The basic idea is as follows: When a piece of wire is stretched, its cross-sectional area is reduced, raising its resistance. Thus if a strain gauge is attached to a steel bridge beam or ship hull, the resistance change indicates the amount of strain in the metal. The amount of bending can then be calculated from the resistance change in the gauge.

Resistance changes are very small (micro-ohms), so very sensitive bridge circuits are used; the signal from the bridge is very small, so proper shielding of wiring is important to avoid noise pickup, which can be larger than the desired signal.

Practical strain gauges are made of many fine wires of metal or foil, usually in a zigzag pattern so that a greater length of wire, and hence a

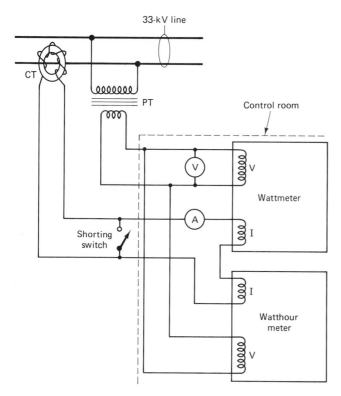

Fig. 3-23 Industrial power metering using current transformers and potential transformers.

greater change in resistance, is obtained. Typical foil gauges are about the size of a postage stamp (see Fig. 3-24).

There are two general types of gauges: the bonded, and the unbonded. The bonded type is usually a foil gauge, glued directly to the structure being tested. The unbonded type uses free wires attached to posts or other supports; the supports are then attached to the structure being tested. The bonded types require special glues and cements which do not cause stress when hardening, and do not flow or crack when stress occurs. Proper application of bonded gauges includes cleaning of surfaces and exact procedures as specified by the manufacturer.

One typical application is in load cells, as indicated in Fig. 3-25; they can be made fairly thin, so that they can be placed under the support legs of a tank, to weigh the contents, or under a weighing scale for trucks. Both of these applications involve wet, dirty operating conditions; the load cells are rugged devices and can be totally sealed so that reliable operation is achieved.

Another application involves measuring the strain in the cable on a large crane used in construction; the strain gauge is attached to a "load measurement assembly" between the hook and the load (see Fig. 3-26).

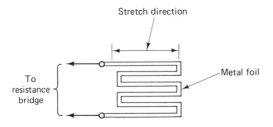

Fig. 3-24   Strain gauge construction.

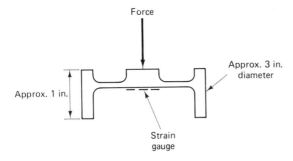

Fig. 3-25   Load cell using a strain gauge.

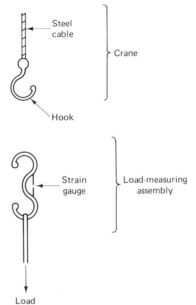

Fig. 3-26 Crane load sensor using a strain gauge.

The load tends to straighten out the measurement assembly and thus stretches the strain gauge.

SENSOR LIMITATIONS

Many measurement techniques seem to work well in a laboratory but are not practical for industrial applications, for a number of reasons. These are discussed briefly next.

Stability. The device may not remain calibrated, and the output cannot be depended on; for example, a pressure sensor may be too sensitive to surrounding temperature, causing false changes in output signals; the electronic circuitry may drift, requiring resetting every few hours; special power supplies may be required, which are inconvenient or expensive.

Linearity. The device may be very sensitive but not have a linear output, as indicated in Fig. 3-27.

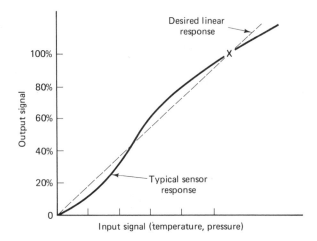

Fig. 3-27   Nonlinear sensor response.

Shock and Vibration.   Vibration is a problem in many installations, and may cause sensor and electronics component difficulties because the device is too delicate.

The examples above are representative of the techniques involved; the reader is referred to the Bibliography for a list of several excellent books that specialize in measurements and transducers.

CONTROLLERS

The controller usually includes the error detector, which compares the desired set point to the actual condition. The error signal is used to produce a signal that is sent to the actuator (which does the actual controlling).

The two most common types of controllers are pneumatic and electronic. Mechanical controllers and hydraulic arrangements (as used on larger power station turbines) are used for special applications. Pneumatic controllers use air pressure and flow in computing devices and electronic controllers use voltage and current. The following discussion applies to both types.

On/Off Control

The common house thermostat is a simple on/off controller. For low room temperature the furnace is turned on and at high room temperature the furnace is turned off. No intermediate steps of "half-heat" are possible.

Note that the thermostat serves as the sensor, error detector, controller, and final control element, all in one device.

This type of control is simple, reliable, and is used where precise temperature control is not required; "cycling" (oscillation) of the temperature is unavoidable, but if that is allowable, the on/off method is adequate, and relatively inexpensive. Typical performance is indicated in Fig. 3-28.

Proportional Control Action

Proportional control is a refinement, used when on/off control produces swings in the controlled variable which are intolerable. Proportional control consists of generating control action according to the amount of error existing. For example, in a temperature control system, the amount of heat demanded would be proportional to the temperature error. In a house heating plant, this would mean controlling the oil burner so that the size of the flame is adjusted automatically according to the size of the temperature error. In industrial work, steam heating systems frequently use proportional control, in which the amount of steam supplied is adjusted according to temperature error.

In an electronic controller, proportional control is accomplished by an amplifier, which multiplies the error voltage by a constant (gain) to obtain the actuator demand signal. Proportional control gives greatly improved performance; temperatures (or other variables) can be maintained closer to the desired set point (see Fig. 3-29). The temperature may still "swing" but over a much smaller range than with on/off control.

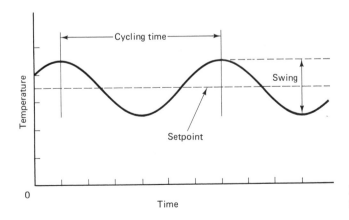

Fig. 3-28 Typical process response using on/off control.

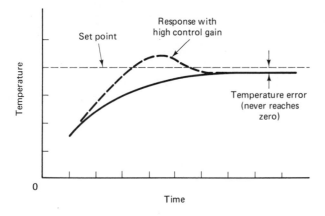

Fig. 3-29 Typical process response using proportional control.

Reset and Rate Control

If precise control is mandatory, requiring very small deviations from set point, the simple proportional control may be inadequate. An additional feature may be added to improve performance: integral (also called reset) action. This addition measures the error and continues to add corrective action in greater and greater amounts until the error disappears (goes to zero) (see Fig. 3-30). Note that more and more actuator movement is demanded even though the error may not be changing for a while. Combined with proportional control, reset action gives excellent control and zero steady-state error. The reset principle is analogous to an integrator, in which increasing output results from a constant input. As long as

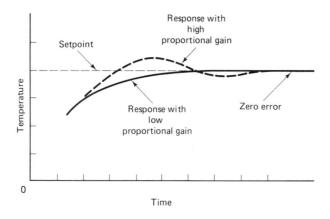

Fig. 3-30 Typical process response using proportional plus integral (reset) control.

error exists, the reset unit will generate greater and greater control action demand.

One familiar example of integrator action is distance traveled in an automobile: Distance traveled is the integral of speed and time. The greater the speed and the longer the time, the greater the distance traveled. At a fixed speed, distance traveled depends on running time. When speed drops to zero, the distance from start position remains fixed. Reset action is easily accomplished with op-amp circuits (see Appendix C). A simplified op-amp circuit for obtaining reset action is shown in Fig. 3-31.

The capacitor must be of very high quality; that is, there must be a very high resistance dielectric (insulating material) between the plates, so that leakage current is almost zero. The capacitor is in the feedback path of the amplifier, so that as long as an input voltage is present, an output voltage is produced and fed back through the capacitor, continuously changing the input voltage at the amplifer; the output voltage thus continues to change. When the input voltage drops to zero, the output voltage holds at its last value, because the capacitor holds its charge.

In the electronic controller, the integral action continues to raise the output signal as long as an error is present (see Fig. 3-32). When the error does reach zero, the output stops rising and holds at the last value (exactly as in the automobile example above: when the speed goes to zero, the car stays where it is, at whatever distance is reached). The output of the electronic controller cannot rise forever, however, because its output cannot exceed the supply voltages to the circuit. Typically, the op-amp will have a ± 15-V supply, so the output usually cannot exceed ± 12 V. Comparing the reset action to proportional action alone (Fig. 3-33), the output of a proportional action unit drops to zero when the error reaches zero.

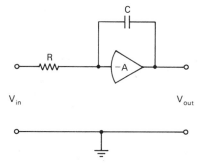

V_{in} V_{out}

Fig. 3-31 Operational amplifier circuit for reset action.

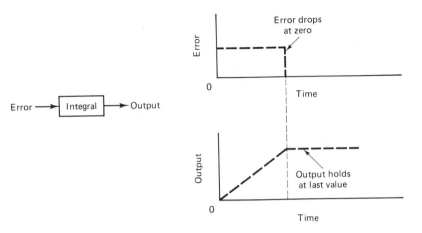

Fig. 3-32 Integral action in a controller with constant error.

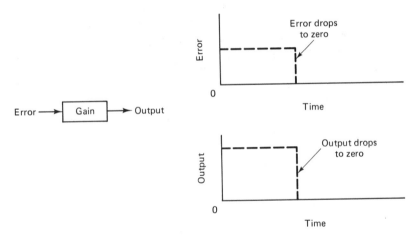

Fig. 3-33 Proportional action in a controller with constant error.

Rate action may be added to produce a faster control system; it acts to cause corrective action depending on the speed of error changes; it senses how fast the error is changing and causes actuator movement accordingly. Rate action speeds up a control system and helps considerably, although not without difficulties (see Fig. 3-34). Rate action can cause problems with cycling of the controlled variable and even cause the whole system to be unstable with larger and larger temperature swings (more about this later).

Rate action may be accomplished with an op-amp circuit such as that shown in Fig. 3-35. Practical circuits may include the shunt resistor

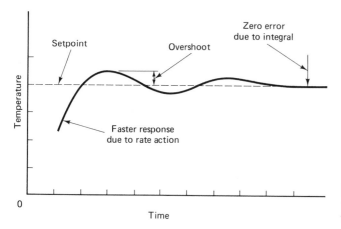

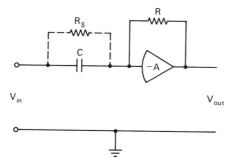

Fig. 3-34 Typical process response using rate action added to proportional and reset action.

Fig. 3-35 Operational amplifier circuit for rate action.

across the capacitor, to produce an approximate rate action, because pure rate action can cause too much response from noise on the input signals. Noise could produce "jitter" on the controller output signal, which would cause the actuator to move erratically.

Dead Band

Up to this point, all control action has been shown to produce an output for any-size error; in practical closed-loop systems it is frequently necessary to include a dead band in the controller settings, which allows a small error to exist before any control action is started. This is particularly useful in fluid flow systems where flow time produces dead time in some readings; in other words, the process changes, but the sensor does not indicate it until some time later. A high-gain setting in a proportional controller will probably produce continuous cycling of the process variable. To stabilize this system (if the dead time cannot be eliminated), a dead band can frequently help. Figure 3-36 illustrates ordinary pro-

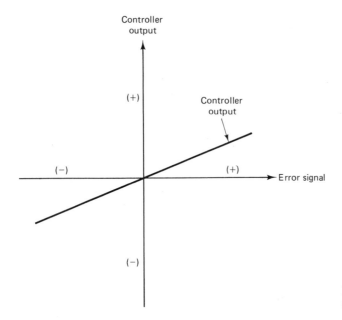

Fig. 3-36   Proportional control action.

portional action; Fig. 3-37 indicates proportional action with a dead band. The dead band keeps the controller from trying to correct small errors and gives the process time to settle out by itself. Control of the process may not be precise, but performance is better than continuous cycling (see Fig. 3-38).

Special Controllers

Various special versions of controllers are used, in which computers are involved. Optimization of a particular variable may be desired, such as maximum production of a given product. Many variables in an entire plant can be controlled automatically to produce a maximum amount of one desired product. Gasoline refineries were among the first to use complex computers to analyze plant performance and make the necessary adjustments automatically.

Ordinary control systems generally use fixed settings of the various controller adjustments. Computing controllers can make calculations and adjust the several plant controller gain, reset, and rate action settings to accomplish the desired results. This field is quite new, but growing rapidly. Digital computers are being applied that can take many data, make complex computations, and decide which variables to change for an optimum output from the plant.

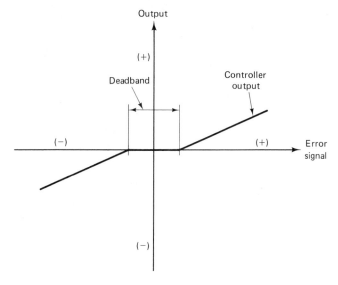

Fig. 3-37 Proportional control action with dead band.

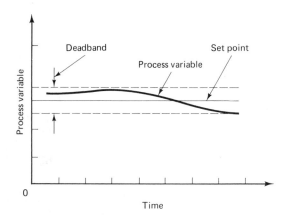

Fig. 3-38 Process response with dead band.

CONTROL ELEMENTS (ACTUATORS)

The control element is the final item in a control loop; it performs the actual work (control function) according to commands from the controller. Typical valve operators are shown in Fig. 3-39.

1. Pneumatic actuators come in several forms: diaphragm types are simple and used for relatively short-stroke applications (see Fig. 3-40).

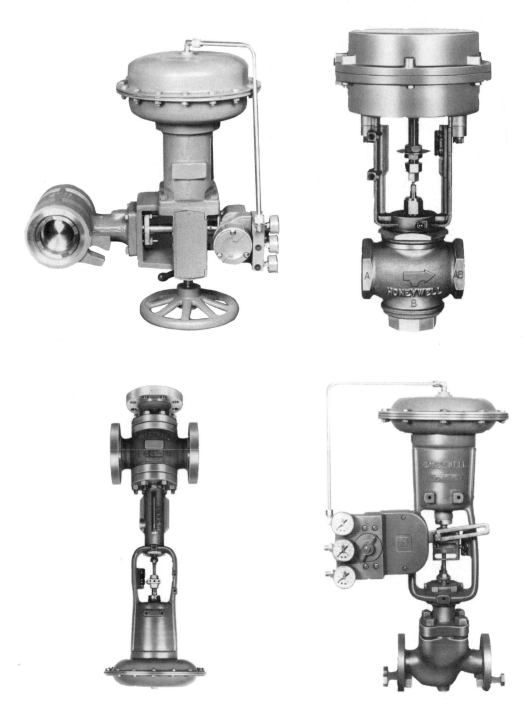

Fig. 3-39 Typical valves and actuators. (Courtesy of Honeywell Industrial Controls Division.)

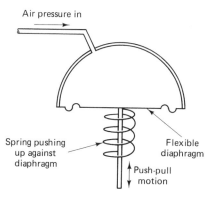

Air pressure in

Spring pushing
up against
diaphragm

Flexible
diaphragm

Push-pull
motion

Fig. 3-40   Actuator using a pneumatic diaphragm.

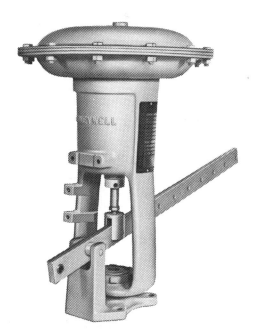

Fig. 3-41   Pneumatic valve actuator. (Courtesy of Honeywell Industrial Controls Division.)

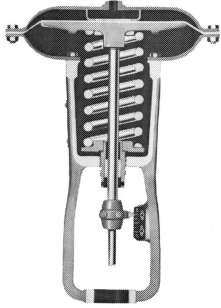

Fig. 3-42   Pneumatic valve actuator: cutaway view showing spring. (Courtesy of Honeywell Industrial Controls Division.)

Air pressure on the top of the flexible diaphragm causes movement downward, against the spring. A typical pneumatic industrial actuator is shown in Fig. 3-41 and a cutaway photograph in Fig. 3-42. A slightly different version is shown in Fig. 3-43 which has the spring over the diaphragm; air pressure under the diaphragm then produces a pull motion, upward.

Cylinder types are used for long stroke requirements, such as furnace damper movement (see Fig. 3-44). Rotary air motors are sometimes used with gearing for long stroke applications, but their complexity tends to reduce reliability.

2. Electric motors with appropriate gearing may be used to obtain movement of valves over considerable ranges. Valve movement proportional to demand may also be obtained (explained later in the chapter).

3. Hydraulic actuators are widely used when high forces are required; high-pressure oil to a hydraulic cylinder can produce large forces to open and close doors or adjust large valves. Hydraulic systems have practically no bounce, as do pneumatic systems due to the compressibil-

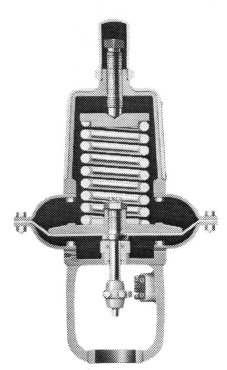

Fig. 3-43 Pneumatic valve actuator: cutaway view showing spring on top of diaphragm for reverse action. (Courtesy of Honeywell Industrial Controls Division.)

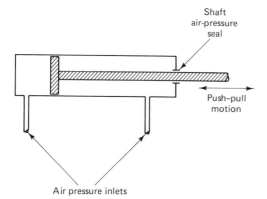

Fig. 3-44   Pneumatic cylinder actuator.

ity of air. Rotary hydraulic motors are also available and have the advantage of large horsepower in small packages and no stall problems (compare to electric motors, which tend to burn out when stalled). Also, in some applications, electricity is unsatisfactory due to the risk of explosion, so hydraulic equipment is a natural. Combinations of electrical and hydraulic equipment is available, providing the best features of each: the speed of electricity and ease of connections (wires), with the safety of hydraulics.

4. Solenoids may be used to move valves, levers, dampers, or other devices which have only two positions; it is a push-pull device, shown in Fig. 3-45.

5. Electrical contacts are a simple means of control; for example, heating elements are turned on or off by a relay.

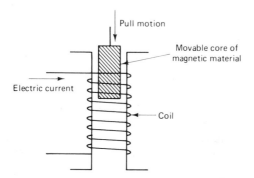

Fig. 3-45   Solenoid actuator.

ACTUATOR DEMAND SIGNALS

Signals are transmitted from the controller to the actuator by any of several methods: electrical voltage, possibly 0 to 10 V dc; or electrical current, 4 to 20 mA; or air signals (pneumatic), 3 to 15 psi. What type of motion results from the demand signals? There are several ways to use the demand signal:

1. On/off control: actuator position either 100% or 0% travel
2. On/off control: actuator velocity either full speed or zero
3. Velocity proportional to demand signal
4. Position proportional to demand signal, open loop
5. Position proportional to demand signal, closed loop

On/off control is the simplest; it can operate two ways: The actuator can be either full open or closed, or the actuator can run full velocity, or zero. In the second case, the actuator stays at its last position when the input command goes to the "off" position. This is a simple system, reliable, and can use three-phase ac induction motors which are constant speed and require little maintenance because there are no brushes, slip rings, or commutator. Air cylinders can also operate in this manner, with full air pressure applied to either side of the piston. Piston speed will be constant with constant air pressure (except for varying loads).

Velocity Proportional to Demand Signal

In this system, an air cylinder could receive air through a valve opened according to the demand signal. Airflow through the valve is controlled by the demand signal and the piston moves at a speed depending on airflow. Electrically, this could be accomplished by a dc motor with field strength controlled by the demand signal; some control circuitry (electronics) is required at the actuator.

Position Proportional to Demand Signal: Open Loop

The air-operated system with a diaphragm actuator operates in this manner, as explained previously (see Fig. 3-40). The diaphragm is pushed up by a spring; incoming air pressure demand signal pushes the diaphragm down until the air force is balanced by the spring opposition.

Position is thus proportional to air pressure except for the possible difficulties caused by the opposing forces of the load; the valve stem may have forces on it, up or down, from the moving fluid in a valve, and thus cause variations in the predicted position.

Electrically, proportional control may be accomplished by using a solenoid pulling against a spring; the current in the solenoid, controlled by the demand signal, determines the push/pull force; some electronic circuitry is required to control the solenoid current according to the low power control signal.

These open-loop actuator drives are not precise, because the movement depends on the opposing forces from the load; 15-psi air pressure might cause full travel sometimes, but only 80% full travel if the load increases later, possibly due to higher flow through a valve or just more friction in the valve stem seals. Still, the open-loop actuator is adequate for many applications and is desirable due to the simplicity.

Position Proportional to Demand Signal:
Closed Loop

If accurate positioning of the actuator is an absolute requirement, a position loop may be arranged as shown in Fig. 3-46. Here the actual position is forced to equal demanded position by using position feedback; a position error signal is generated in the actuator loop controller, which may even have reset action. This loop is called a positioner, because it

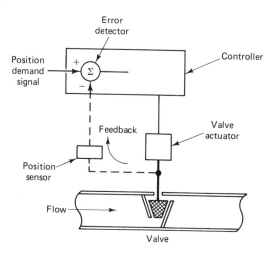

Fig. 3-46 Pneumatic valve positioner with position feedback loop.

Valve stem

Slotted lever feeds
actual valve position
to error detector in
the enclosure on the
left.

Fig. 3-47 Pneumatic valve positioner. (Courtesy of Honeywell Industrial Controls Division.)

causes the actuator to move to a demanded position, with little or no
error, compared to the demanded position, despite load variations. An
industrial valve actuator with a positioner is shown in Fig. 3-47. The
position sensor and controller is mounted on the left side of the valve.
This figure shows an air-operated valve controlled by a current loop position signal, 4 to 20 mA. The actual valve position is sensed by the slotted lever, which connects inside the box to the position error detector.
Air pressure to the diaphragm operator is adjusted automatically so that
the valve stem position is exactly proportional to the position demand
signal, despite changing loads due to flow through the valve, or friction
in the valve stem packing. Note that this valve positioner loop is a loop
within the main process control loop.

WIRE AND CABLE

Why a section on wire? Because there is quite a choice available when
selecting wire to connect control equipment. The simplest thing to do is
run single wire stranded; however, this is not wise because of the like-

lihood of electrical noise pickup. Twisted-pair shielded is the next higher quality, but this is available in many types: single-braided shield, double-braided shield, braid with drain wire, and so on. The big differences are in *coverage* of the shield, which indicates how much electrical noise can get in through the shields. The highest quality is foil-wrapped pairs, with drain wire; this type has extremely good specifications (high coverage) and is probably the best type of *data-carrying* wire. Cables can be purchased with many pairs, each individually foil wrapped, with a shield over all. In some cases, *crosstalk* can be troublesome: Signals in one wire, or pair, are picked up in adjacent wires, causing false signals. The cable with separately shielded pairs would minimize that possibility. Some equipment may require coaxial cable; even triaxial cable (two separate braided shields) may be required to reduce problems with electrical noise pickup and ground loops; the two shields can be connected to different "grounds" where necessary.

Each type of shielding requires special procedures to connect to terminal blocks: what to do with the braid, the foil wrap, the drain wire, and so on. Each cable vendor provides adequate instructions for proper installation to connectors and terminal blocks to obtain the best performance (resistance to noise pickup). It is vital to follow these instructions, because otherwise the best noise immunity might not be obtained.

Use the best cable you can obtain, because that will help avoid future equipment malfunctions due from noise pickup in the cabling. In addition, it is generally wise to run all signal cables in metal conduits for additional electrical shielding as well as mechanical protection. A good discussion of cable choices and shielding effectiveness may be found in Appendix C.

SUMMARY

Hardware for closed-loop control systems must be selected carefully, for ruggedness and stability as well as accuracy; proper installation of wiring is important, to avoid electrical noise pickup, which can ruin the performance of an otherwise good system.

STUDY QUESTIONS

1. Why should sensor signals be run as high-level signals?
2. Explain six types of voltage-generating transducers.

3. Sketch and explain the operation of hydraulic actuators.
4. What is a valve positioner loop?
5. Describe two versions of on/off control.
6. What problems can occur with poorly shielded cable?
7. Why are cable ground connections important?

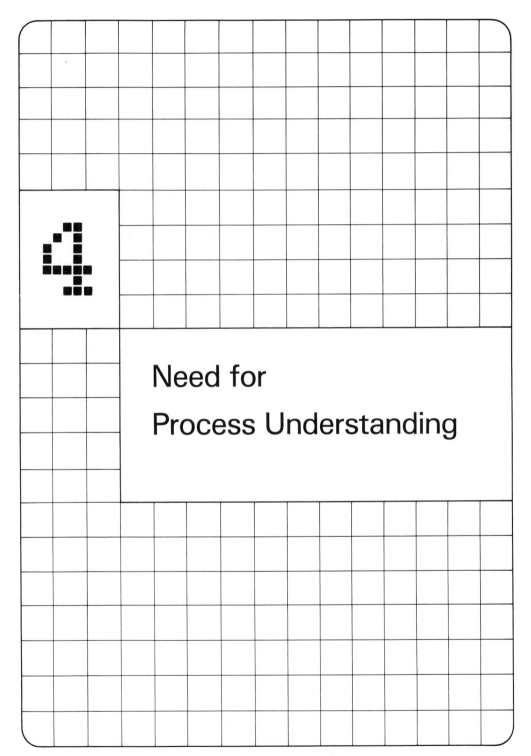

4

Need for

Process Understanding

OBJECTIVES

To emphasize the need for cooperation between process designer, the control designer, and technicians

To explain some of the difficulties that can occur in closed-loop systems

To outline the probable effects of excessive dead time in closed-loop systems

To indicate the probable results of placing sensors too far from the variable to control

To indicate problems that can occur when using sensors with long time constants

In this chapter we discuss some of the difficulties involved in specifying the pieces (hardware) of a control system. As a technician, you must understand what the control engineer needs; frequently, you can point out difficulties that have occurred with existing control equipment, and assist in writing better procurement specifications in order to avoid similar troubles with new equipment.

It is truly surprising to discover how many control systems are put together by selecting measuring devices, controllers, and valve actuators from catalogs *before* really analyzing the overall system. Then after installation the control adjustments are "fiddled" in an attempt to get satisfactory system behavior. This is often called "design," but it certainly is not the best approach—and depends on luck and good technicans to obtain satisfactory system performance after installation. Sometimes the whole system will not perform as intended, and then major changes are required to modify a control system which was not logically designed in the first place. Even the process or machine may require modifications. There is likely to be much expense and time lost with this approach. In this chapter we point out some of the considerations in choosing the control components.

Sometimes the machine designers do not understand what particular details can make control easy or difficult. They design the process or machine, then tell the control engineer to "design controls." In many cases control is difficult or even impossible, because of the way a process is built; sometimes even a small change in piping arrangements would make better control possible.

The process designer will certainly be willing to consider modifications to an initial design, especially if far superior control is possible and process throughput much higher, resulting in greater revenue. Close

cooperation between the process designers and control personnel will reduce the number of control problems to be solved after the whole system
is started up. The important point is that the process must be understood
and described in mathematical terms by both the process designer and
the control engineer, before attempting to select pieces for a control system. It is not wise to guess at a valve speed or an amplifier gain, for
example, until some basis is available for choosing such numbers. That
basis is an analysis of the system to be controlled.

DESIGN EXAMPLE: MIXING VAT

As an example of control system development, consider the mixing vat
shown in Fig. 4-1. This is a frequently used scheme for heating and mixing chemicals, food, or paint. Two or more pipes supply material to the
vat, and the mixture leaves through the outlet pipe. The mixture must
be heated because it is too stiff when cold, but should not be overheated
because it will either ruin the batch or create a fire hazard. Temperature

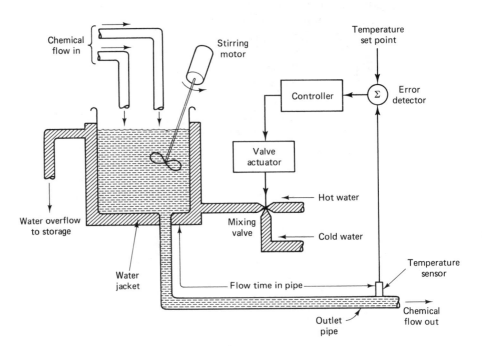

Fig. 4-1   Temperature-controlled mixing vat.

control is therefore required. The vat shown is heated by a water jacket around the vat; the water temperature depends on the position of the valve that mixes hot and cold water going to the jacket (some chemical mixtures give off heat; hence cold water may be required at times.) A stirring motor is shown to ensure that all the incoming materials are thoroughly mixed and that the mixture is at the same temperature throughout.

The temperature control system consists of a temperature sensor (for example, a resistance thermometer or thermocouple), a controller, a valve, and a valve actuator. Selection of these items from a catalog requires information such as that outlined in the following paragraphs.

Resistance Thermometer or Thermocouple

What temperature range, speed of response, and accuracy are required? Will a 20-second response be adequate, or is a 2-second response absolutely required? Response is dependent on construction, and a thermometer in a heavy enclosure will respond very slowly. Response time is generally rated by "time constant," exactly as stated for an electrical circuit with resistance and capacitance (it is the time required for the output signal to reach 63% of its final value, for a step change in the input temperature—and this temperature is the temperature of the mixture). The response may not be a perfect time constant as shown in Fig. 4-2, but probably will be more like that shown in Fig. 4-3; note that in Fig. 4-2 the output signal starts up immediately, whereas in Fig. 4-3 there is some lag before the output signal starts rising. The difference is due to

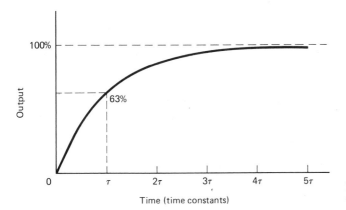

Fig. 4-2 Response to step input: one time constant.

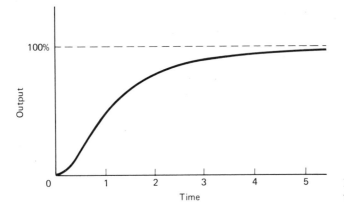

Fig. 4-3 Response to step input: two time constants.

the several parts, chiefly the wall thickness of the assembly and the mass of the resistance element itself. The largest portion of the time constant is usually due to the size and wall thickness of the well, together with the size of the vat and water jacket.

Accuracy.   The accuracy required depends on the original system requirements; does the mixture have to stay with 5 degrees of the desired temperature or is 10 degrees good enough? Perhaps only 1-degree variation is allowable for some chemical mixtures. How fast will the set point be changed? Will the set point be changed frequently? Will there be upsets in the process due to cold material being dumped in?

Given a choice, a control engineer would probably prefer the most accurate and fastest thermometer available. This is not wrong, technically, but the faster item may cost 10 times the other. Also, the faster thermometer may be more delicate and not reliable; hence the choice is not always simple.

Placement of Sensor.   This is a vital consideration, often forgotten until too late. It does little good to select a 2-second thermometer, for example, and then place it in the pipe far away from the vat it is supposed to measure. The flow time of the mixture in the pipe may be 5 seconds or more (dead time, from vat to thermometer) and the thermometer indication will always be late, simply because of its location downstream. Good control may be impossible with excessive dead time. Note that in Fig. 4-1 the thermometer is shown far down the outlet pipe.

Figure 4-4 indicates the temperature sensor response; note that there is an appreciable time before the sensor shows any change at all. A better location would be in the vat, so that the actual mixture tem-

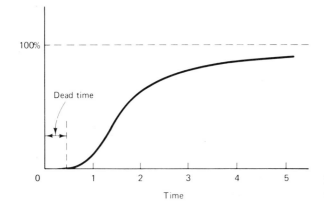

100%

Dead time

0   1   2   3   4   5
                 Time

Fig. 4-4 Response to step input: two time constants plus dead time.

perature is measured. If that location is impossible, then the thermometer should be as close as possible to the vat. That fact should be made clear to the persons responsible for vat piping layout and installation.

Controller

Here the engineer has many choices: pneumatic, hydraulic, electronic, or even mechanical devices. Also, decisions must be made regarding speed of response and the range of control adjustments. These adjustments determine how much valve movement results from a 1-degree temperature error (controller gain, in other words). In addition, integral and rate action may be required and their settings must be specified.

The range of control adjustments must be specified also: whether a gain range of 0.5 to 4.0, 5 to 20, or 50 to 200, or whatever. The proper range must be determined before ordering a controller. The point is that if a certain controller setting of 2.5 is required, a controller with a range of 0 to 2000 would be nearly impossible to set properly. Unless specified, a standard catalog controller will probably have too wide an adjustment range or the wrong range (more on this in Chapter 11).

Valve Actuator Speed

Selection of valve speed is a critical item; if the speed is too low, poor temperature control will result (the temperature of the vat contents will vary far from the desired value before correction). If the valve speed is too high, poor control may also result because the valve will go rapidly from all hot to all cold water in the water jacket, causing wide swings (oscillations) in the mixture temperature. Valve speed must be chosen with regard to overall system behavior, preferably by computer simulation studies (discussed in Chapter 5).

None of the choices noted above should be made until the process itself is understood (here the process is the vat with the water jacket). "To understand" means to know much about the process, such as the following items:

1. How fast does the mixture temperature change when the jacket temperature changes? This depends on how much water is in the jacket, how many pounds of mixture is in the vat, and the thickness of the vat walls.
2. How fast is the mixture flowing out of the vat? Does the flow rate vary?
3. How much do the temperatures of the inlet materials vary?
4. Does the amount of material in the vat vary? If so, the response time will vary, complicating the job of specifying controller settings.

Some of this information is easy to obtain; other numbers are not so easy. Sometimes the engineer must resort to actual experiments on the vat itself to determine performance and speed of response. In other cases, enough can be predicted with an analysis on paper, perhaps using a computer to help solve the equations.

SUMMARY

The point is that the characteristics of the process must be known before selection of control components can be done properly. Otherwise, there is the possible risk of failure to obtain good control, which in addition to being embarrassing and expensive, causing delays in plant operation, might even be dangerous if the vat contents are temperature sensitive and explosive. Particularly important is the cooperation of the process designer, who really knows more about the process than anyone else and who can contribute much to a successful control design.

STUDY QUESTIONS

1. Explain why the control engineer and technicians should be consulted when a new process is being designed.
2. Why is sensor placement so important?

3. What would be the effect on system response if a very slow sensor were installed?

4. What data on the process does a control engineer need?

5. What would be the effect on loop response if the process response were faster than expected?

6. What type of information does the control engineer need to supply to the process designer?

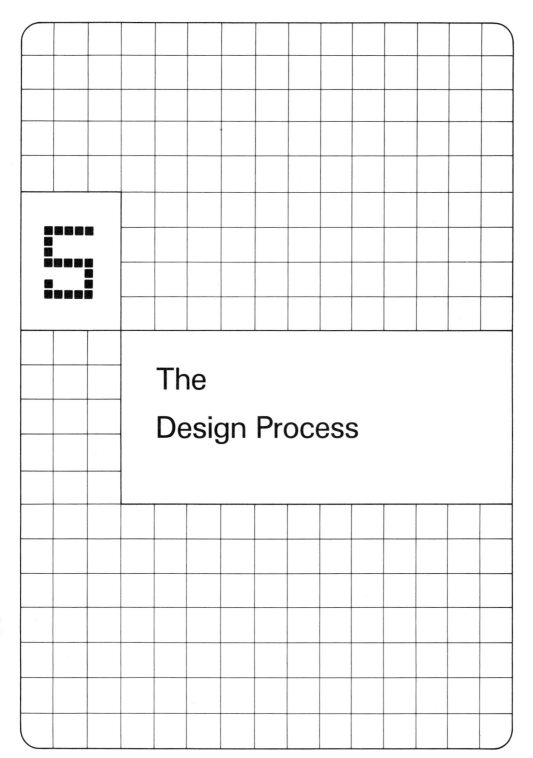

The

Design Process

OBJECTIVES

To explain the steps followed by an engineer in developing a well-engineered system

To point out the value of a simulation study

To outline the need for interface equipment

To anticipate problems with noise pickup

To emphasize the need for a comprehensive failure analysis

To discuss the possibility of plant runaway

Control system design should follow a logical sequence, so that when plant startup time is reached, controls will operate properly. Some control decisions, such as those involving sensor time constants, placement of sensors, and valve speeds, are difficult to change once the items are installed. It is far better to avoid unpleasant surprises by proceeding carefully through the design process.

It is important to consider the possibilities of accidents due to equipment malfunctions and human error. A safety system should be designed to operate separately from normal controls. There are, therefore, two control systems: one for normal closed-loop operations and the second for safety, monitoring plant conditions.

A well-engineered system will proceed through the following steps:

Analysis ←──────────────┐
Design   │
Simulation   │
Specification   │
Testing and evaluation │
  (repeat if necessary) ─┘
Safety analysis
Safety system design
Safety system specification

ANALYSIS

Analysis is the study of the overall task; it will include determining the characteristics of the process, whether it is a simple water-heated vat or a nuclear electrical power plant. Analysis will also include study of the available control equipment: sensors, actuators for valves, and controllers, which typically will be electronic or pneumatic. Note that the term "available" is used; this is important for several reasons; if very special requirements are specified for a device, custom design may be required; custom design is expensive and should be avoided if at all possible. The other reason is reliability; custom-designed items have not been proven and may be unreliable. In addition, shipping dates may be missed because the manufacturer has trouble meeting specifications. It is important to use standard catalog items unless there is a good technical reason to do otherwise.

Another vital consideration is standardization of equipment in a plant; if most of the control components are similar, the technician will be familiar with troubleshooting and can do a better job keeping the equipment on-line. Spare parts will be less of a problem, and parts may even be "borrowed" from another control system in the plant, when that other system is not running at the time a breakdown occurs on a vital process controller. This standardization may mean that all equipment will come from one controls manufacturer; of course, that is risky, to be "locked in" to one supplier, for several reasons:

1. Prices may be increased at any time, particularly if the vendor knows that you have no alternative.
2. Spare parts may be discontinued at any time, leaving you with the choice of finding equivalent parts or buying new equipment when you cannot repair the old.

It is thus highly desirable, if at all possible, to have at least two equipment suppliers for everything.

Analysis will include writing down the characteristics of the process in mathematical form, if possible. Sometimes this can be done only approximately, but the approximation may be accurate enough to predict process behavior. In this case the engineer must estimate the accuracy of the equations and make allowances in the design procedure—so that if the process behaves somewhat differently than expected, the controls can be adjusted to cover the situation.

DESIGN

Once the process is well understood and described mathematically by the engineer, the selection of control components can begin. Many estimates will be made regarding speed of response required for the thermometers, flow meters, pressure sensors, and so on. Here the experience of the engineer is valuable, for good judgment can be used to estimate what will work, with technical requirements balanced by economic considerations (these decisions will be evaluated by the next step, simulation).

SIMULATION

Simulation generally refers to the use of computers for representing mathematical equations. When the equations of the process, sensors, controller, and actuators are put in the computer, the engineer has a working model of the entire system and can study its behavior and observe whether initial estimates of time constants, speeds, and so on, were sufficiently accurate. Numbers can be changed and the effects observed. This is a very valuable tool for the engineer and is the way that nuclear plants have been studied—*before* they are built. Also, operating crews for ships and nuclear plants have been trained on simulators before the actual plant is finished. In this way, all (or most, hopefully) of the mistakes can be made before the actual plant is involved. An actual full-size control panel can be used. The operators can get a "feel" for plant behavior and learn how best to control it: what set points are critical, what dangerous behavior to avoid, and how to control it manually if required.

In this simulation phase, the engineer can determine the effects of a slower thermometer, for example, or a different valve size or valve speed. The effect of different process response times may be investigated, to cover the situation that may occur: The process may perform slightly differently than expected! The effects of plant flow times can be studied, together with placement of sensors on piping, which results in dead time. The best combination of controller settings can be found, so that when the actual process is run, the controllers can be set very nearly correctly the first time—rather than fiddling for several days. Time on the computer is less expensive than actual time on a big plant, not to mention the price of mistakes. During the design of the plant, the simulation studies can be of great assistance in predicting behavior and avoiding the pitfall of inadequate control. The control designer should therefore work closely with the process designer, to reach a plant design that is controllable.

There is always the possibility of instability of the control loop (process plus controls). Instability is the condition in which a process variable oscillates continuously and *no* controller settings can be found to achieve steady control. Factors that can cause instability are as follows:

1. Excessive dead time (flow time) before sensors.
2. Excessively slow thermometer elements (long time constants).

3. Unrealistic specifications; it may be impossible to achieve a 0.1-degree temperature control, for example, with the vat as designed; the walls may be too thick; or the water jacket may be too large (long time constant).

Thus instability may be due to the design of the process itself; *no* control system can overcome some basic faults. The process is part of the control loop, and the control engineer can suggest what possible changes would help make control feasible (or possible!). It may be that the desired control accuracy cannot be achieved with the proposed vat design; in that case the simulation study would help locate the problems, and help the process designer make changes that will make satisfactory control possible.

SPECIFICATION

After much study of the computer model, the engineer is ready to write specifications for the various components involved: speeds, accuracies, time constants, range of control adjustments, and other items. The specifications will include items such as operating temperature and humidity requirements, power supply requirements, maintenance provisions, and the like. Note that the engineer is specifying what will work and not just blindly selecting items from catalogs; it may be that standard catalog items will be adequate, but specifications must go beyond the usual sales material. (See Chapter 11 for further comments on writing specifications.)

TESTING

The testing phase is vital; each item purchased by the specifications must be tested to see that it does meet the specifications. Often this verification is omitted—but it is unwise because there are, unfortunately, vendors who will ship equipment that does not meet all specifications, hoping to get by. (That is the reason that many experienced engineers test and inspect everything, 100%, when received at the loading dock.) During this testing phase, many data (which may not be in the instruction books with the equipment!) can be collected on each control component, such as:

Controller adjustments: actual adjustment range of each knob
Actuator speeds and response times

Sensitivity of equipment to line voltage and frequency changes

Equipment calibration procedures and difficulties (and test equipment required)

Drift in equipment calibration

Equipment sensitivity to vibration

Noise output of equipment, both audible and electrical

Errors in instruction books: where the equipment does not agree with the book descriptions.

Once each item is tested, the components can be assembled, installed, and the entire closed-loop system tested. This phase will require much cooperation between the process designers (chemical engineers, nuclear engineers, mechanical engineers, or whoever) and the control designer. The safety of the process is foremost; therefore, much testing will be done at low power or slow speed until the controls are demonstrated to be functioning properly.

During this testing and evaluation phase, the control engineer will make many measurements to see if the control system performs as intended and keeps temperature or pressure, or other variable, within specified bounds. If not, the simulation study should be repeated, to find out why actual results differ from simulation results. Perhaps one or several equations need to be modified to make the simulation agree with the actual process. The engineer can then suggest new controller settings for the actual process controllers and obtain better overall performance.

The foregoing sequence of events indicates the procedure for large projects such as nuclear power plants and chemical refineries. Smaller projects may not include simulation on computers, due to the expense. In that case, the engineer must do some analysis on paper to solve the equations and predict system response. There are many techniques for this, but none are as convenient as simulation on a computer.

INTERFACE EQUIPMENT CONSIDERATIONS

The signals from sensors may require modification, filtering, level shifting, or impedance transformation, depending on what signal the sensor sends and what the control room electronics requires. Output signals may also require modification to air pressure signals or current loop signals. These functions, generally called *signal conditioning*, can include the following items:

1. Analog-to-digital and digital-to-analog converters
2. Current-to-pressure and pressure-to-current converters
3. Pressure-to-voltage converters
4. Filters to keep out radio and electromagnetic interference
5. Frequency-to-voltage converters
6. Digital signal conversion from various formats: binary-coded decimal to binary, for example
7. Data-line surge suppressors

Filters have already been discussed for power lines, which are supposed to keep line noise out of the control equipment. The data lines should have the same protection; regular digital signals are typically 0 or 5 V, for logic systems. Pulses and spikes may be picked up from many sources, possibly causing damage to semiconductor devices. It is thus always good practice to specify surge suppressors on each data line in and out.

Note both *in* and *out*; pulses can travel in both directions on a signal line, not just in the same direction as the desired signal. Thus a large pulse from a nearby radio transmitter may be picked up on an output wire to an actuator, but that pulse may travel back into the output circuitry and cause damage.

Location. The signal-conditioning equipment will generally be located in the control room, both for convenience (calibration and maintenance) and for the controlled environment (temperature and humidity).

NOISE ON ANALOG SIGNALS

Frequently, analog data from remote sensors will arrive at the control room polluted with noise of several varieties. There is always 60-Hz noise from power lines; other frequencies may be picked up from control room video monitors (15,750 Hz or higher) and switching-type power supplies (20 to 300 khz); Radio receiver local oscillators (both AM radios and FM radios) radiate strong signals; even the control room computing equipment radiates a variety of high-frequency noise, due to the fast pulses used.

The first idea may simply be to connect capacitors across the data lines to ground; that may reduce the undesired noise, but will also slow down response of signals from the sensors. The benefit of a fast sensor could be completely eliminated with those capacitors. A fast steam pressure sensor, for example, could appear to respond very slowly.

A better approach would be to install low-pass filters in the data lines, or single-frequency traps, if only one frequency is troublesome. The low-pass filter would be designed to pass those frequencies that are necessary for good response of the transducer and overall system (the simulation study could indicate what frequency response is necessary). The single-frequency trap could be designed to keep out a specific signal, such as 60 Hz or 15,750 Hz.

NOISE GENERATED BY ELECTRONIC EQUIPMENT

A source of noise frequently overlooked is the actual electronic equipment in the control system. Switching power supplies, chopper-stabilized amplifiers, and computing equipment can generate strong signals over a wide range of frequencies. Proper design of each unit should minimize the amount of noise that gets out on power lines and data lines.

SAFETY CONSIDERATIONS

In most cases, safety systems should be entirely separate from normal process measurements and controls. For example, separate temperature sensors should be installed, operating from a separate power line, which is battery backed and not the same power line that supplies normal control equipment.

It is all too possible to have some type of failure that will disable both the normal control channels and the over-temperature trip system. The designer must make every effort possible to keep the safety functions separate, so that plant shutdown is always possible, and fast enough to prevent damage to machinery or hazards to personnel.

Frequently, the safety system consists simply of a limit switch on the process recorders; the moving ink pen writing on the chart is supposed to operate the switch if the process variable rises above a certain value. But what if the normal temperature measurement system does not indicate properly, and reads low, or fails? Then no safety system exists!

During the design process, the engineer and technicians (and everyone else involved) should always keep in mind the consequences of possible accidents, whether due to human error or to equipment malfunction. Processes utilizing high-powered machinery or inflammable materials are especially susceptible to accidents. What happens when line power fails? When the computer fails? When an interface channel fails?

Malfunctions may be considered in three categories:

1. Normal problems
2. Operator errors
3. Equipment problems

Normal problems are those difficulties that are expected, such as conveyor jams, tool breakage in automatic machines, web breakage in paper mills, utility-power-line failure, and the like. These types of problems occur often enough to be considered routine, and generally follow some pattern. They are annoying but are generally expected and are not too serious. The operators and maintenance people know what to do, and get the plant back on-line relatively quickly.

Operator errors occur when someone fails to do something, or does something incorrectly, in the wrong sequence, or with the wrong timing. Some of these types of errors will be repeated, and perhaps the control panels should be modified so that any possible operator confusion is avoided. Perhaps a different type of display instrument or better labeling of controls would make them less prone to improper operation. Perhaps better training of operators would help.

Equipment problems are unexpected breakdowns in the process itself or in the control equipment. Every imaginable type of failure should be considered, along with consequences. The consequences are the vital concern: Would a valve actuator failure cause leakage of poisonous material? Would a pump failure lead to a boiler explosion? Would a temperature sensor failure lead to dangerous over-temperature conditions? Would a relay sticking closed (with welded contacts) cause an actuator to run continuously until a mechnical stop is reached? Would a cable break cause a controller to open a fuel valve 100%? These types of questions are very difficult to answer, but that difficulty is not a reason to ignore them. Every effort should be made to estimate the results of every conceivable equipment failure, and then plans made to avoid plant damage or personnel hazards.

Whatever the reason for malfunctions, there must be a safety system, that (1) alerts the operator to the problem, and (2) shuts down the appropriate machine if dangerous conditions are approached. However, too many automatic shutdowns are a nuisance and plant production (revenue) suffers. Thus the operator should receive warnings in sufficient time to correct the situation if at all possible. If an automatic control is faulty, perhaps the operator can run a pump on manual control for a time so as to keep the rest of the plant on-line.

The designer must also remember that when an automatic safety system causes too many shutdowns, there will be a tendency for operators (and shift supervisors!) to defeat (bypass) some safety circuits with a jumper wire, force relays closed with toothpicks, or prop valves open with sticks, to keep the process running. Of course, these types of defeats can lead to disasters. Therefore, the control designer must not install too many automatic trip-outs, but just enough to ensure safety. That is a difficult balance to determine!

PLANT RUNAWAY

Plant runaway is a situation in which a control failure causes operating conditions to reach dangerous levels very quickly; for example, a steam boiler fuel valve runs to full open and stops there. Boiler temperature, steam pressure, and boiler-drum water level (and other variables) then become incorrect and their control loops try to make corrections. However, so many things go wrong, so rapidly, that the operator cannot decide what to do.

Closed-loop controls can cause plant run away, as mentioned in Chapter 1. Any failure that produces an incorrect error signal will cause an incorrect actuator demand; if the error signal remains constant (due to a fault) the actuator will run until full travel is reached, either full open or full closed.

One of the main hazards of these types of failures is that the operator probably has little experience in handling unusual emergency situations. Normal plant operation may allow the operator to become complacent and not alert. A strange sequence of events (due to control failure) will cause confusion because the operator cannot determine quickly what caused several variables all to "go bad" seemingly all at once. What should the operator do first?

In a steam boiler plant, for example, a wide-open fuel valve would lead to very high steam pressure, along with low drum water level, in a very short time. A steam safety valve would, hopefully, open and keep pressure within reasonable limits. The operator might panic and turn off everything, including the manual fuel oil shutoff (which is probably the correct thing to do).

It would then take some time to determine what caused the improper pressure conditions, especially if the fault that started everything has gone away (intermittent failure!) after the manual emergency shutdown.

HOW RUNAWAY CAN HAPPEN

In a closed-loop control system runaway can be caused by any failure that causes an actuator to move incorrectly; anything in the loop that causes a faulty error signal can lead to runaway. In addition, an actuator can fail, resulting in full travel.

Some possible control system faults are as follows, not necessarily in the order of probability.

1. Open or shorted sensors.
2. Cables, plugs, and sockets: shorts, opens, or intermittents due to poor soldering, corrosion, broken wires, or bare wires.
3. Cables damaged by machinery or vehicles.
4. Human errors: disconnecting or unplugging cables while equipment is active.
5. Electronic equipment failures (internal).
6. False signal pickup in sensor cables: electrical noise, such as from electrical welding, radar signals, or radio signals (even from portable radio transceivers which were unwisely placed inside an equipment cabinet).
7. Power supply faults in control equipment: balanced plus and minus dc voltage supplies that fail, for example, leaving one side energized and the other dead; such an unbalanced situation will cause operational amplifier circuitry to output a maximum signal (saturate) in one direction or the other. Also, radio-frequency noise can originate in switching-type dc power supplies. That noise can travel on the dc supply lines, and be rectified in a diode junction somewhere, causing false error signals.

It is thus obvious that all possible failures cannot be predicted and avoided. The only way to avoid plant damage and/or hazardous conditions is to provide a safety system as a watchdog on normal control-loop operation.

Error Signal Monitors

In a closed-loop control system, the size of the error signal can be monitored. Normally, the error signal will be small, so an excessive error signal could be used to initiate an alarm to alert the operator that something should be watched; if the excessive error persists, the operator could change to manual control until the reason for excessive error is found.

Refer to Fig. 5-1; the safety system could monitor the error signal and cause an alarm if error reaches level (1), and automatic shutdown if the error magnitude reaches level (2). Thus the operator may have time to correct plant conditions to avoid the shutdown. Alternatively, the safety system could monitor a process variable such as temperature, then alarm at one level and automatically cause a shutdown at a higher level (see Fig. 5-2).

Power Source Monitors

Especially important are power supply monitors. Each control system supply, especially the balanced plus and minus sources, should have a monitor that automatically disconnects the automatic control functions if voltages are far out of tolerance or unbalanced. For example, the monitor on the ± 15-V supply should trip (and cause a return to manual operation) if either voltage drops below 13 or rises above 17, or if the two have a difference of more than 2 V (exact numbers for these trip settings should be determined when the control system is first tested by itself). There would probably not be time for operator action, so quick trip-out to manual control would be advisable.

Similarly, monitors on supply air pressure or hydraulic pressure to actuators are important. If pressure falls slowly, the operator should be alerted, so that an orderly shutdown can be started. If pressure falls suddenly (possibly due to a pipe break), an automatic shutdown is probably advisable.

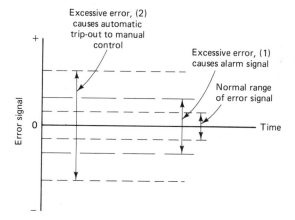

Fig. 5-1 Error signal monitor: trip levels.

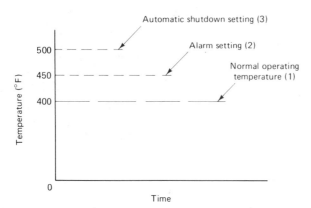

Fig. 5-2 Process monitor: trip levels.

Safety System Backup

For very high energy systems such as utility electrical generating plants, even a third safety level is provided (backup), completely separate from the first two:

1. Normal closed-loop controls
2. Safety system
3. Safety system backup

This third level is the "last ditch" type of safety control, not expected to be used, but provided to prevent damage (or minimize damage) if the first two systems do not operate properly. Examples of level 3 items are as follows:

1. Mechanical steam pressure safety valves on boilers

2. Blow-out panels on boilers or chemical reactors: these panels are built intentionally as relatively weak sections, so that any explosions will rupture the panel, releasing pressure and minimizing damage to the main structure. These panels may require some time to replace, but that is a small price to pay if major damage is avoided. The size of these panels would be based on experience, because there is probably no way to calculate accurately the pressure due to undesired explosions.

3. Frangible disks are metal disks built to break (rupture) at predicted pressures. They would be sized to open at pressures above the safety valve settings, for example, so that if the safety valve opened and

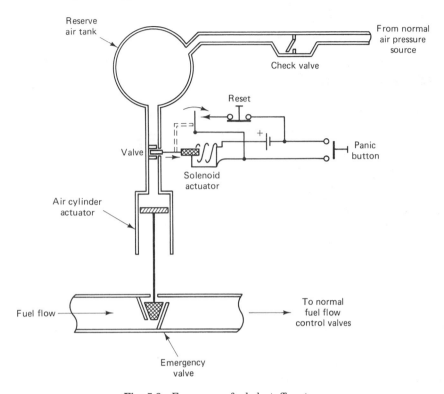

Fig. 5-3 Emergency fuel shutoff system.

pressure was still too high, the frangible disk would rupture and provide another flow passage to reduce pressure. These, too, would take some time to replace, but are somewhat more convenient than blowout panels.

4. Emergency valves operated by reserve air tanks or hydraulic storage tanks. These tanks should be piped separate from the main air supply or at least piped through check valves, so that the reserve pressure is not lost if the main source fails. The valves would be independent valves normally open which could be driven closed very quickly with air or oil pressure. An oil-burning steam boiler might have such a valve on the fuel line. For a panic shutdown, as indicated in Fig. 5-3, the operator could push a panic button that causes the emergency valve to close very quickly (of course, the reserve air tank should have a pressure sensor which causes an alarm if the pressure drops below that which is required for an emergency shutdown). Note that the solenoid actuator has a lock-in contact so that the emergency shutoff valve is held closed until the reset button is operated, and the emergency shutoff valve is manually reopened.

SUMMARY

A careful, logical approach to control system design will result in a control system that adequately controls the process. Comprehensive testing of each item, before installation, will aid in later maintenance and troubleshooting.

In addition, safety systems should be provided, completely independent of normal control loops and functions. The safety system is intended to monitor plant operations and alert the operator when variables are out of tolerance and to shut the process down if dangerous conditions are approached. If the process is potentially hazardous, a third level of safety equipment should be provided to at least minimize damage and hazards to personnel.

STUDY QUESTIONS

1. List five steps in the design process.
2. What might happen if some design steps were omitted?
3. Explain why it is desirable to use standard catalog components and equipment.
4. What is simulation?
5. Explain several advantages of a simulation study.
6. What is plant runaway?
7. What could cause runaway?
8. Why are three levels of control and safety systems sometimes advisable?

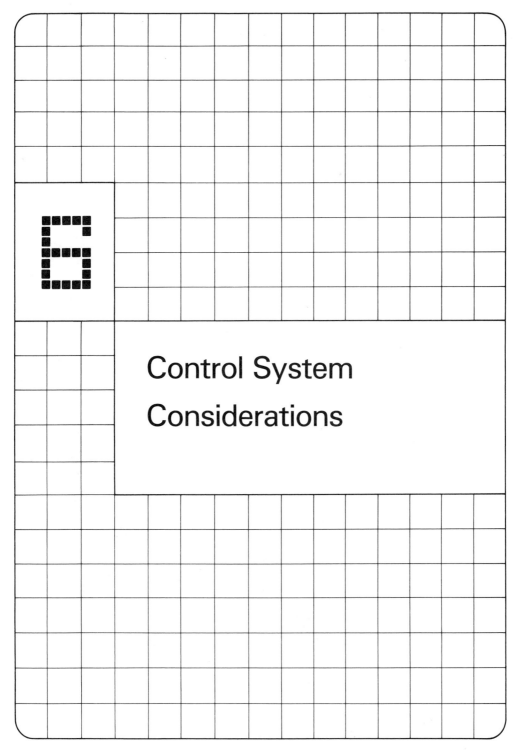

6

Control System Considerations

OBJECTIVES

To indicate some important factors to consider when planning a new or revised control system

To outline some power line and grounding problems

To outline procedures that can improve reliability and safety

To indicate typical closed-loop tune-up procedures

To emphasize the importance of telephone communications

The topics discussed in this chapter can be considered *applications engineering*; some of these topics are frequently overlooked but are vital for reliable, trouble-free installation and operation of control systems.

A primary objective of the engineering department is to provide a control system that is convenient to operate, reliable, and easy to maintain. The topics discussed in this chapter are very important, involving plant designers, installers, operating personnel, control system designers, and maintenance personnel. Maintenance personnel may be the last to be consulted, but that is a mistake: people who are involved with day-to-day problems can certainly contribute good ideas when control improvements are planned.

LOCATION

The control room should be located in a convenient location: not, for example, on the fourth floor, because technicians will be going frequently between the control room and the remotely located control components.

The control equipment should be in a fireproof structure so that minor fires outside do not spread and destroy the control room. An interesting problem may arise: Local fire codes may require a sprinkler system in the control room; there is then the possibility of false actuation, and water damage due to the sprinkler system. That possibility should be eliminated if at all possible. Safe fire extinguishers, of the carbon dioxide or halon types, should be provided so that minor fires can be extinguished without causing more damage than the fire.

Where will the control equipment be located? Control consoles, equipment cabinets, cable racks and conduits, and emergency power supplies must be in convenient locations, with adequate space and suitable air conditioning.

SPACE

The size of the control room is determined not only by the control console and equipment cabinets, but by considerations of how many people will be working there. Normal conditions may call for two operators, but during troubleshooting, two or more technicians, two supervisors, and vendor representatives may be present. Also, test equipment such as oscilloscopes, signal generators, and extra chart recorders may be rolled in. Sufficient space must therefore be provided to allow these people to accomplish their job. Both front and rear access to all equipment cabinets and console should be provided, because the plant operator may be required to maintain plant operation while the technician works on some minor problem.

Space should also be provided for the inevitable "tours" by visitors to the plant. It should be possible for visitors to walk through without disrupting the operators or interfering with the maintenance personnel.

RADIO SIGNAL INTERFERENCE

More often than one would think, there may be problems with interference from radio signals. There may be a commercial AM or FM transmitter nearby, or a transmitter used for communication around the industrial plant (portable radio transceivers used by the electronic technicians, for example) which can induce signals into the control wiring and cause mysterious malfunctions. Even radio signals from aircraft overhead can cause problems.

Radar transmitters are particularly bad in this regard, because they transmit very high power pulses. If your factory or plant is located in one of the new "industrial parks" near an airport, an aircraft surveillance radar will probably be nearby, and can induce large pulses in the control wiring every time the radar antenna sweeps around. If this is the case, good shielding of all electronic equipment and wiring will be a wise precaution. Even then, you should realize that equipment cabinets will have the doors open much of the time, for routine maintenance, measurements, and troubleshooting, so if radio signal interference is likely, the entire control room may require shielding (screening in the walls, floors, and ceilings).

ENVIRONMENT

The control room should have a controlled environment, for temperature, humidity, and also contaminants such as dust particles and fumes. Electronic equipment is very durable, particularly with integrated-circuit technology, but if temperatures are maintained at reasonable levels, say below 100°F inside the cabinets, reliability is greatly improved. Humidity must be controlled also, to avoid moisture condensation on printed circuit boards and connections. Dust is a real hazard if disk drives are used in the computer systems. Very small particles can cause a disasterous "crash" of the disk system and loss of stored information.

Corrosive fumes are frequently present; sulfur dioxide, hydrogen sulfide, and other fumes can cause corrosion of contacts in cable connectors and integrated-circuit sockets, eventually resulting in faulty contacts and intermittent operation. To minimize problems with dust and fumes, air-conditioning systems can be designed for a positive pressure in the control room, so that all air leakage is outward from the control room; the air intake to the system then can be filtered and the air quality controlled much better than if air reaches the control room from any open door. The air intake should be located very carefully, so that outside fumes and smoke will not be picked up and pumped into the control room. Audible noise in the control room should be minimized; equipment fans sometimes create quite an annoying whine, and air-conditioning systems often are noisier than necessary. It is thus obvious that the control room should have its own air-conditioning system rather than simply being tied into an existing building duct system.

INSTALLATION

Control equipment, including the control console and equipment cabinets, should be designed for maintenance personnel as well as the control operator; both front access and rear access should be provided, because the maintenance technician must often make measurements inside the equipment, from the rear, while the control operator continues working. Cables, test points, and so on, must be accessible while the plant is operating, without interfering with the operator.

POWER SOURCES

Normally, the control equipment will be operated from standard ac lines; however, a separate circuit should be provided for maximum reliability. That line should come from a separate step-down transformer, as close to the utility high-voltage source as possible. The objective is to reduce to a minimum the number of circuit breakers, fuses, and switches between the utility source and the control room—to reduce the risk of power disconnections due to unintentional operation of switches in the plant. That separate power line should be carried in a conduit and routed so that the possibility of physical damage is minimal. Underground service is preferred to avoid physical damage from machinery, and also damage from fires near the line.

It would be very desirable to have two separate electric utility lines to the control room, from different utility sources; these lines should arrive at the control room by different routes across the plant. It would also be highly desirable to have an automatic switchover between these two power lines, if one fails.

Battery backup of control system power is preferred also, both for emergency lighting in the control room and as power for vital control functions. Utility power may disappear for milliseconds, minutes, or hours. It is practically impossible to provide backup power for the entire plant, but safety considerations will indicate which machinery and which control functions must be maintained, despite utility power interruptions. A standby diesel generator can be provided, but that unit will require several minutes, possibly, to start and pick up the load. Battery-backed power for control functions is still preferred to avoid any interruption.

LINE-FREQUENCY VARIATIONS

A different type of problem may occur when using standby diesel-electric generators: The voltage and frequency regulation is not nearly as good as that of the electric utility. Some electronic equipment (sensors and controllers) may use resonant power transformers to regulate voltage; these transformers operate properly only at the design frequency, usually 60 Hz in the United States. If the standby electric generator swings from 55 Hz to 65 Hz, for example, these resonant power transformers may cause excessive voltage swings due to the frequency changes. If a power supply is specified to maintain a dc output voltage to 5 V despite load variations and line voltage changes, it may operate correctly until the line frequency changes, which may cause the output voltage to reach 7 V. When writing specifications (more on this in Chapter 11), it is wise to specify that resonant power transformers shall not be used, or that all specifications shall be met over the specified line frequency and voltage range.

OTHER POWER-LINE PROBLEMS

Even when a separate power line to the control equipment is installed, several other problems can arise: voltage dips, brown-outs, load shedding, and spikes.

Voltage dips (sometimes called "sags") occur on utility lines when large power loads are switched on or off (in your plant or in the utility system). The normal line voltage may drop to 80 V or less, for milliseconds or up to several seconds.

Brownouts are intentional voltage reductions by the utility; that is done to prevent overloading of generating facilities when unexpected system loads occur. A generating plant may stop producing, requiring the remaining generators to pick up the load. A transformer or transmission line may fail, leaving a portion of the utility network overloaded. The utility system dispatcher could order a temporary voltage reduction, which has the effect of reducing system load until additional generating capacity can be brought on line.

The intent is to avoid shutting down electric generating stations due to overload, because that might lead to overloading the remaining facilities and result in a domino effect in which the entire utility network collapses (this happened on the east coast of the United States several years ago).

During a brownout, line voltage may be reduced to 100 V from the normal 120 V, which in turn reduces the amount of power supplied. Such a brownout can last several minutes to several hours, until the utility system can recover to normal conditions.

Load shedding is another possibility; this is a planned procedure, during which some utility customers are totally disconnected, to avoid utility system overloading. Naturally, no one is in favor of that, so perhaps brownouts are less troublesome than total power failure, except that brownouts are very hard on electric motors, causing overheating due to excessive current.

Voltage spikes are frequently present on utility lines, due to normal utility switching activities, lightning, or welding machine operation. These spikes may occur randomly or repetitively on the sine-wave power. For example, lightning may produce a series of spikes at random, as shown in Fig. 6-1. Lightning strikes do not have to be in the immediate vicinity to be troublesome—even if 20 miles away, when lightning strikes a power line, spikes may travel on the transmission lines and cause very high voltage pulses to appear on your 240/120-V control room line.

Welding machines may produce repetitive spikes that appear on the normal sine-wave line voltage, as shown in Fig. 6-2; note that the spikes appear regularly, in phase with the line frequency and may shift phase when welding equipment changes current output.

Spikes may be very narrow, down to nanoseconds in width, and are very difficult to observe. An average oscilloscope (response to 15 MHz) will not show such narrow spikes. Lightning spikes, particularly, will not show (unless you are lucky!) because they are not repetitive. Even

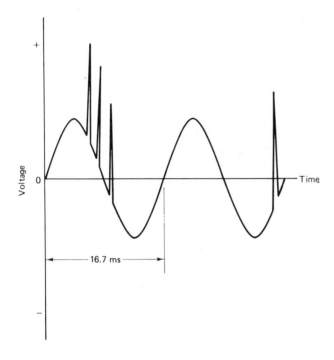

Fig. 6-1 Random spikes on power line (60-Hz power line assumed).

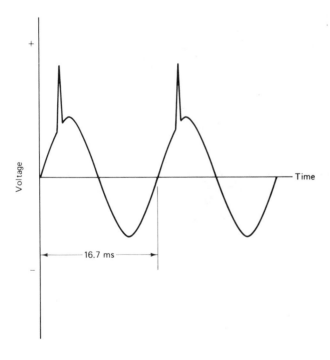

Fig. 6-2 Regular spikes on power line (60-Hz power line assumed).

oscilloscopes with response to 60 MHz may not show such random spikes, because there is nothing to synchronize the oscilloscope sweep.

Worse yet, the spikes may occur only occasionally, causing malfunctions in the control system. Even if you have an excellent oscilloscope and a camera, the spikes may not be there after you get set up, so you cannot prove that spikes caused any problems.

The best defense against spikes is good power supply filtering in the electronic equipment cabinets (after bringing in the power lines through separate conduits, underground if possible).

INTERNAL EQUIPMENT PROBLEMS

An interesting (and troublesome) problem has been observed when using simple power supplies with a typical circuit shown in Fig. 6-3. This is a bridge rectifier circuit, which may feed an excellent voltage regulator and produce smooth dc, well regulated, under normal conditions.

What is not shown is the stray capacitance between the primary and secondary windings. The secondary side thus receives voltage two ways: inductively, as intended, and capacitively. The capacitive coupling can be troublesome, because it can make incoming spikes worse. An approximate equivalent circuit is shown in Fig. 6-4. The RC (resistor–capacitor) circuit produces pulses which are steeper (faster) than the incoming pulses. It is a differentiator, which puts out a voltage proportional to the rate of change of the input voltage. In this instance, the input pulse occurs from A to B, not too large, but the output voltage D to E is very fast and of higher voltage than the input pulses. Actually, the output pulse is two pulses, because the output of the RC circuit depends on the slope of the input pulse; the first part of the output pulse is generated when the input pulse goes from A to B (positive slope); the pulse EF is generated when the input pulse goes from B to C (negative slope).

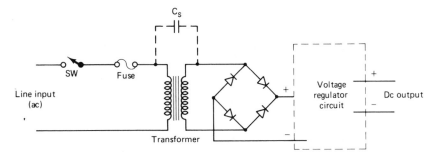

Fig. 6-3 Power supply with stray capacitance across transformer.

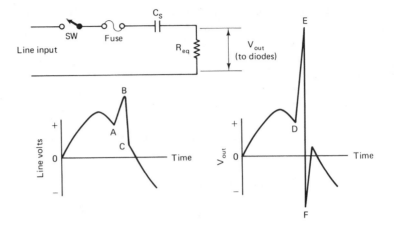

Fig. 6-4 Power transformer: equivalent circuit and response to voltage spikes.

Thus the stray capacitance actually makes spikes appear worse and very difficult to filter out on the secondary side. Those secondary circuit pulses can destroy rectifier diodes and integrated-circuit regulator circuits by exceeding the rated peak inverse voltage (PIV) limits.

Electrostatic Shielding in Transformers

The cure for this problem is electrostatic shielding between the primary and secondary of the power transformer, which allows only magnetic coupling between primary and secondary. That shielding consists of a wide copper band between the primary and secondary. Pulses are bypassed to ground instead of being coupled to the secondary side (see Fig. 6-5).

Transformers built in this way are somewhat more expensive, but certainly worth it, to avoid random problems from suspected spikes (inasmuch as it is nearly impossible to prove that power-line spikes are causing problems; better to do everything feasible to avoid the possibility).

Line-Voltage Regulators

One preventive measure is to install a line-voltage regulator to supply the electronic equipment in the control room. Good regulators can correct for voltage dips and brownouts and filter out spikes and electrical

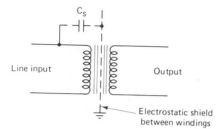

Fig. 6-5 Power transformer with electrostatic shielding.

noise arriving on the power line. Of the many models available, select one that is not frequency sensitive, in case a standby electric generator is used (which may have an output frequency varying from 55 to 65 Hz, or worse).

Also note that it is very difficult to test these regulators yourself. For example, there is no simple, inexpensive way to set up a test source to insert spikes on the power line to see if the regulator eliminates them. The only real choice you have is to purchase equipment from reputable, experienced vendors who can be trusted to meet their own specifications.

CABLING FOR CONTROL SIGNALS

Control signals to and from the control room should be run in conduits separate from power lines; even then, these conduits should be kept separate, at as great a distance as possible from high-current circuits, because it is possible to induce currents in the metal conduits, which in turn will induce voltages in the control wiring inside the conduit. If wiring must be run only in cable racks and troughs, install these as far away from power wiring as possible.

Ground Loops

When planning the system cabling, keep in mind that ground loops must be avoided. This means that cable shields should be terminated separately at equipment terminal blocks, not just all soldered together and connected to the nearest pipe or cabinet ground terminal. By terminating each shield separately, engineers can connect only one end of the cable shield to ground, and avoid ground loops. Also, during plant startup and troubleshooting, it is frequently necessary to change the first plans for grounding shields; if each shield is separately terminated at terminal blocks, changes in ground connections will not be too difficult. Note that if all input/output circuits are equipped with optical isolators, then problems with ground loops are greatly reduced (more on this in Chapter 11 and Appendix B).

Grounding Problems

Proper handling of shields and grounds is very important when very low voltage signals are to be transmitted, such as from strain gauge transducers. Sometimes double-shielded cables or coaxial lines must be used, when very low voltage signals are to be transmitted, such as from strain gauges or transducers. Ground connections become very complicated, because you have two separate shields around the signal wires, and each shield may be connected to a different "ground" or one end left unconnected. There are at least three grounds to consider:

1. *Building ground.* This is the frame of a steel building; hopefully, all of the structure is bonded together but still will not be at the same potential, and not the same potential as a water pipe ground.
2. *Cabinet grounds:* the electronic equipment cabinets.
3. *Electronic equipment chassis grounds,* within a cabinet.

What should be done with the cable grounds? That is a difficult question and the engineering department must arrange these grounds so that troublesome currents do not flow in shields and induce false voltages in the inside cable wires. As an example, consider Fig. 6-6; two equipment cabinets are shown, with a coaxial signal wire between. If the shield is connected to the cabinets at both ends, there is the possibility of current flowing through the shield if grounds G1 and G2 are not at the same potential. Figure 6-7 shows a rough equivalent circuit; inductive coupling (transformer action) can induce false voltages in the inside signal wire, possibly causing false sensor signals or false actuator

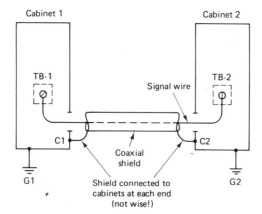

Fig. 6-6 Signal cable illustrating ground loop through cable shield.

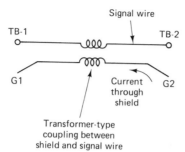

Fig. 6-7 Equivalent circuit of ground loop.

movements. To avoid that problem, it would be better to ground the shield only on one end, at G2 or G1, but not both.

At the same time, "floating grounds" are to be avoided because of shock hazard to personnel (a floating ground is a separate power line return from an isolation transformer, for example, which is not connected to a building frame ground such as indicated in Fig. 6-8). Beware of power distribution schemes that do not have the green safety wire and all equipment cabinets connected to a good earth ground! The wiring shown might reduce noise pickup in the control equipment, but it is not safe because the green wire, on the right, is not connected to a good ground through the green safety wire, on the left side. The floating ground may allow the cabinets to be several volts above ground and be dangerous to touch. The best philosophy is: Safety is the first consideration and electrical noise reduction, second.

A side note: *Never* use test equipment plugged into an isolation

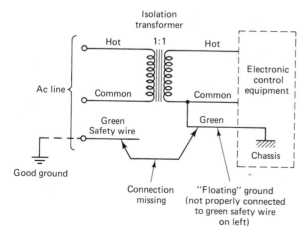

Fig. 6-8 Circuit illustrating unsafe use of isolation transformer.

transformer. Such a connection leaves the test equipment cabinets floating at an unknown voltage, with no real "ground."

REMOTE EQUIPMENT

Some electronic equipment will be located far from the main control panels. Many types of sensors require complex electronic equipment which must be located close to the actual sensor. That equipment must be installed with much care, because of the following problems:

1. *Environment.* Heat, dirt, vibration, humidity (wind and rain).
2. *Power source.* Where will the remote equipment be connected? Where will test equipment be plugged in?
3. *Electrical noise.* Due to adjacent electrical switchgear and machinery.
4. *Communications.*

All the electronic equipment should be mounted in metal cabinets, where possible, to keep out dirt and moisture as well as radio interference. If these cabinets are outside, some roof overhead should be provided to protect both personnel and temporary test equipment, so that adjustments and maintenance can be done safely when it is raining, or worse. Electronic components should not be mounted in the same cabinets with electromagnetic relays or contactors which produce sparks; those sparks represent energy that can induce false pulses into the control equipment.

What about operating power for sensors and actuators at these remote locations? It is desirable to run a specially designated power line from the control room to the outlying electronics. Dc (direct current) power such as 48 V dc would be preferred because it can be run with the signal and data lines in the same conduit. Each electronic unit could then incorporate a voltage regulator to provide whatever voltages are required internally. However, this low-voltage dc system may not be practical, especially if equipment from several vendors is used, each requiring 120 V ac.

If ac power is required, a specially designated "instrument power" 120-V ac line should be installed all over the plant, so that reliability is maximized. Do not just plug in the remote electronic equipment to the nearest 120-V ac circuit!

This suggested system is similar to hospital systems which have especially marked circuits connected to a high-reliability source, battery backed, to operating rooms and other vital areas where power failure would not be tolerated.

COMMUNICATIONS

Telephone communications are generally provided in the control room, but how do you talk to a technician working on a pressure sensor? During troubleshooting, reliable communication must be provided from the control room to every location where control components are installed.

Of course, portable walkie-talkies can usually provide adequate communication, but there may be situations where radio use is not suitable, such as the following:

1. *Hazardous locations* (flammable substances nearby); radio transmitters can produce sparks.
2. *Shielded areas*, such as steel buildings, where a radio may not work reliably.
3. *Sensitive areas*, such as near low-level interface racks or computer cabinets, where a radio transmitter may introduce false signals, causing more trouble. (Perhaps a prominent sign should be posted: "Do not operate radio transmitters in this area.")

For these areas, telephone communication is preferred. A separate battery-powered system, separate from the normal dial system, should be provided, with plug-in jacks at each remote equipment location. Locations that are frequently visited could have a telephone permanently installed so that the technician does not have to carry one there. These telephones need not be on a dial system but only connect to the control room, selected with patch cords, at a patch panel (a party-line system is not a good idea, because if that line were damaged in one area of the plant, the whole phone network could be disabled). The technician could use a walkie-talkie (before entering the sensitive area) to request control room personnel to connect the appropriate telephone line.

For extreme reliability and a backup to the first system, sound-powered phones could be provided to allow communication when all electric power fails. Whether or not these are justified depends on the nature of the process and the hazards and risks involved in losing communication.

FAIL-SAFE?

Always keep in mind that the plant must keep operating because downtime is expensive. Reliability of control equipment is very important and thus ease of maintenance is vital. Always ask the question: What happens when the ac power fails or a control signal to an actuator disappears?

Failures do occur because nothing is 100% reliable. All control components should be designed to fail-safe, if at all possible. That means, for example, that a valve actuator failure should result in closing a fuel oil valve, not opening it 100%. Other control components may not be as easy to specify, but every effort should be made to estimate what might fail and what might happen. Perhaps fail-safe on some valves might mean to open completely; others would be safer if the valve stays where it is when the failure occurs. The best answer depends on the process involved.

If at all possible, all control loops should have some backup method of operation. If a loop controller is out of service, manual operation should be possible (and convenient).

CONTROL LOOP STARTUP

How do you start up a closed-loop system? That question does not have a simple answer because the answer depends on the process being controlled. Many processes, such as small boilers, will be started with manual control, then switched to automatic closed loop when conditions are correct. Other systems can be started in closed-loop operation and allowed to settle out by themselves (the water bath temperature control system for photographic solutions, for example).

At the time of startup of any system, the error signal in the control loop will probably be large; if the error is connected to the controller, large signals will be sent to the actuator; the process may then respond with fast changes and excessive overshoot before settling down (see Fig. 6-9). If the overshoot is not acceptable, the actuator can be controlled manually by the operator, bringing the process variable up slowly to the desired operating conditions, or nearly so, then switching to automatic, as indicated in Fig. 6-10.

During startup, process variables usually behave somewhat differently than during normal operation. A controller gain setting for normal operation might *cause* intolerable and dangerous transients during startup.

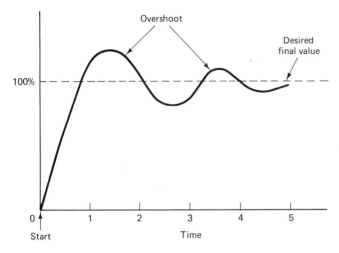

Fig. 6-9 Process overshoot on startup.

Of course, it is theoretically possible to program a computer to control the set points (demand signals) to a loop and raise the set point slowly as would be done manually. However, if many loops are involved, it is very difficult to predict exactly what sequence and how much and how fast each loop set point should be raised. Probably human input will be *vital* and the only way to start up complex plant operations. A human being can modify any planned procedures if something does not respond as expected, whereas it is nearly impossible to create a computer program with all possible process behavior and situations predicted.

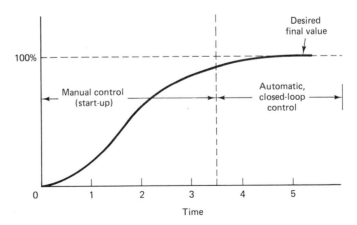

Fig. 6-10 Process response with manual startup.

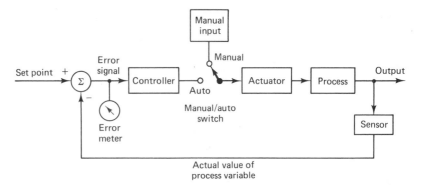

Fig. 6-11   Closed-loop block diagram showing manual/auto transfer switch.

Generally, when starting up a process control loop, the manual/automatic switch should be on "manual" if a large error exists—unless you know that the process can stand large transients. To avoid these transients a meter should be connected to read the error, as shown in Fig. 6-11. Perhaps one operating procedure would be to reduce the set point to obtain a small error, switch to automatic, then raise the set point carefully to the proper value, watching the process response.

What is desired is "bumpless" transfer from manual to automatic, meaning no large jump ("bump") in process output data shown on the chart records. Many commercial controllers have a bumpless transfer feature, but different vendors accomplish this in different ways. Therefore, you must understand exactly how your equipment operates. Probably, startup will still be a problem, requiring some care before closing the automatic control loop. Generally, it is undesirable to close any control loop if a large error signal is present.

LOOP TUNE-UP

After the loop is closed, the controller must be adjusted so that the process variable is properly controlled. The controller will probably have several adjustments: gain, integral action (reset), derivative action (rate); this type of controller is generally called a PID controller (proportional, integral, derivative). A provision for dead band may also be provided (as described in Chapter 3).

These adjustments determine the overall loop response to disturbances; there are two primary types of disturbances that the controller must handle:

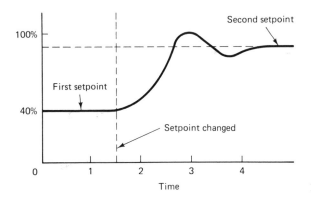

Fig. 6-12 System response to set-point change.

1. Changes in set point
2. Load disturbances in the process

The first occurs when a set point is shifted to a new value and the second when something outside the control loop causes process changes (in a boiler, water temperature of feedwater, or a steam flow change due to throttle changes, for example). In the first situation, the controller should bring the process variable up to the new set point quickly and smoothly with little or no overshoot, as shown in Fig. 6-12. This figure indicates overshoot and oscillation before settling down.

In the second case, the controller is supposed to bring the process variable back to the set point, when something changes the process output. Controller gain and reset and rate actions must be adjusted (tuned) for adequate response, which generally means fast response with little or no overshoot, and no continuous oscillations. Tune-up will be different for the two types of disturbances indicated above, because loop behavior depends on *where* the disturbance comes into the loop. In the first case the set point changes, so it represents a disturbance before the controller. In the second case, something external to the loop changes process output, so the disturbance comes into the loop after the controller. A decision must be made regarding which situation is most important, then tuning the controller to perform best on that one, with perhaps poorer response on the other.

GENERAL GUIDELINES TO TUNE-UP

Suppose the set-point change is the most important: first, set the controller reset and rate actions to zero; set the gain to a predicted value (from the engineering department simulation studies), or if no setting is

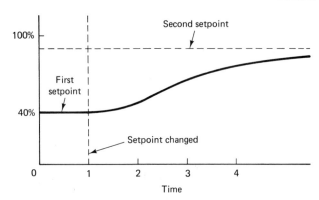

Fig. 6-13 Sluggish loop response with low proportional gain.

available, set gain to a small value. Change the set point slightly and observe process response. If process response is sluggish, as shown in Fig. 6-13, raise gain slightly until some overshoot occurs when the set point is changed, then reduce gain until overshoot disappears (sometimes a slight overshoot is acceptable.)

Next, observe how close to the second set point the process variable comes; raise the reset action slightly and observe that the process variable is corrected and rises up to the new set point (see Chapter 3 for a discussion of reset action in a controller).

Proportional action (control gain setting) will bring the process variable close but not exactly to the set point. There will always be some error remaining with proportional action alone (see Fig. 6-14).

Reset action will bring the error to zero because the controller demand to the actuator will continue to change until error is zero. Thus it is called reset action because it resets the variable to the set point exactly, as indicated in Fig. 6-15.

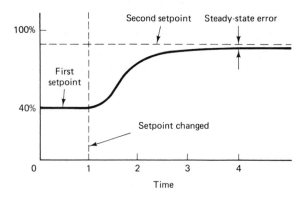

Fig. 6-14 Closed-loop response with proportional control action.

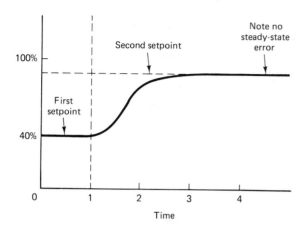

Fig. 6-15 Good closed-loop response with proportional and reset action.

If you keep raising the reset adjustment, the process variable will begin to show overshoot; reduce the proportional gain slightly so that overshoot is minimized and error is reduced to zero in a reasonable time.

If fast response is required, you can raise the rate action, which speeds up the first part of the response curve. However, this action can cause oscillation and unstable behavior, so adjust slowly and carefully (see Fig. 6-16). It will probably be necessary to reduce proportional gain even more, when raising rate action, to avoid excessive overshoot.

This is a trial-and-error approach, in which proportional gain is raised first, then reset action is brought in to reduce long-term error. Next, rate action may be added to speed up response. These adjustments interact with each other, and system response depends on the process characteristics, as well as controller settings. If the process has been

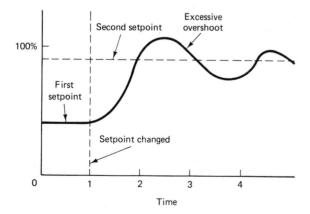

Fig. 6-16 Closed-loop response with excessive rate action.

studied using computer simulation, some controller settings will be specified in advance as a starting point for tune-up. If no simulation has been done, tune-up can be done as outlined above.

Always keep in mind that too much rate or reset action will cause overshoots and even continuous oscillations, so if that behavior is observed on the process variable, reduce rate action first, then reset, and see if behavior improves.

RECORD KEEPING

When equipment is installed, it is very wise to keep good records of cable routing and cable termination information. Each wire should have a wire marker attached, corresponding to installation drawings. Troubleshooting is much easier if wiring is marked and agrees with the drawings! Later, during plant startup, all controller dial setting should be recorded, so that later you can determine whether some equipment fault occurred or whether someone improperly changed a controller gain setting.

SUMMARY

After the analysis is done, there are many practical considerations to consider before a good control system installation is achieved. Control equipment is sensitive to environment, which includes, among other things, temperature, humidity, air pollution (fumes and solid matter), and vibration. In addition, power-line sources are extremely important. Equipment must be installed and shielded properly, including considerations of possible ground loops and noise pickup. Wiring and cabling is a vital part also, because poorly planned wiring can ruin the performance of an otherwise excellent control system. Extensive shielding of equipment and wiring may be required if radio interference is troublesome.

STUDY QUESTIONS

1. List several types of electrical noise that might interfere with control signals.
2. What type of interference do electronic welding controls cause?

3. What would be the symptoms of radar signal pickup?

4. Explain some difficulties that might be caused by brownouts and voltage dips.

5. What is utility "load shedding"?

6. Outline several reasons for good record keeping.

7. Explain how you could tune up a control loop using proportional and reset control actions.

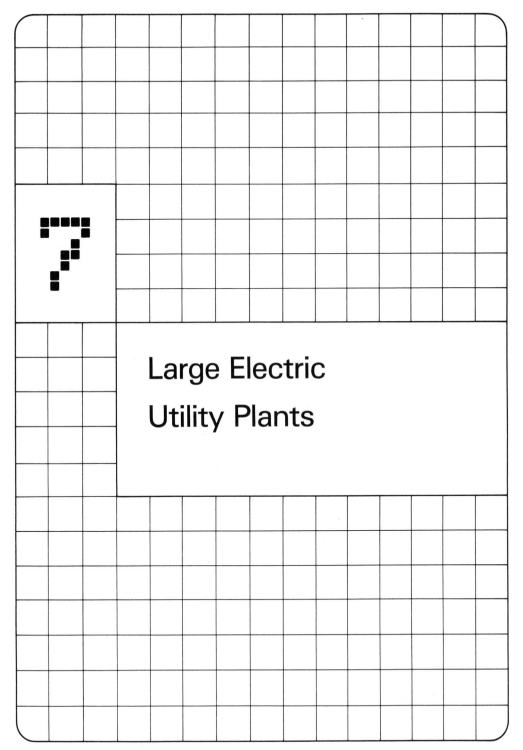

7

Large Electric Utility Plants

OBJECTIVES

To describe construction and operation of a typical electrical utility power plant

To outline control requirements and typical control loops

To indicate some operational characteristics and problems

To describe some auxiliary system requirements

To describe some safety concerns

To discuss communication requirements of an electric utility system

Electric utility companies build and operate very large electric generating plants; these may generate from 800 to 1300 MW in each boiler-turbine-generator combination (and there may be several of these on one site). The overall principles are the same as discussed in Chapter 2 (small boilers), but steam pressures and steam temperatures are much higher for the sake of higher efficiency. Also, steam is returned to the boiler for reheating, sometimes twice, after passing through the turbine. Feedwater and air for combustion are usually preheated using stack gases.

Fig. 7-1 The Chesterfield Power Station is the largest fossil-fueled power station in the state of Virginia and one of the largest in the Virginia Power system. Its generating units have a capability of 1,298,000 kW. The station is located on the James River about 15 miles southeast of Richmond, Virginia, in Chesterfield County. (Courtesy of Virginia Power.)

Figure 7-1 shows a modern power station owned and operated by Virginia Power, near Richmond, Virginia.

CONSTRUCTION

A general sketch of a coal-fired electric generating plant is shown in Fig. 7-2; the operating cycle begins with the feedwater pumps at the lower right, pumping water through pipes (called "tubes") in the furnace walls (water walls). This serves two purposes: preheating the water for efficiency, and cooling the walls. After leaving the tubes in the walls, the feedwater reaches the lower drum ("mud drum"). Steam generated in the furnace tubes rises into the steam drum, from which it flows through the superheater tube sections. The superheat process adds energy to the steam, raising its temperature considerably. As collected in the steam drum the steam is barely steam and is quite wet (saturated steam). By superheating, the steam is made considerably hotter and at the same time, drier, which is better for the turbine. The interior of a large boiler ("steam generator") is shown in Fig. 7-3.

The turbine is usually built in several sections, actually being several turbines on the same shaft. The superheated steam first passes through the high-pressure turbine and is then routed back to the reheat furnace sections, before returning to the next sections of the turbine. If the plant uses "double reheat," the steam may again be routed back to the furnace before reaching the third section of the turbine.

By reheating the steam, additional energy is extracted from the hot furnace gases, which otherwise would simply go up the stack, representing wasted energy. The less heat up the stack, the more efficient the overall plant will be. Additional efficiency measures include air preheating, which uses stack gases to preheat combustion air so that a hotter fire is obtained.

From the last turbine section the steam flows to the condenser, where it is condensed back to water, and stored in the sump. The feedwater pump draws water from the sump to begin the cycle over again.

A coal-fired plant also includes coal pulverizers, which grind the coal to a fine powder so that it will burn very rapidly in the firebox. Forced-draft blowers push the burning coal and hot gases through the furnace chambers toward the stack; the induced-draft fans assist in moving the gases up the stack. These fans are driven by very large motors, up to 5000 hp.

Figure 7-4 shows a modern turbine-generator room; note the size of the man next to the turbine. The generator (alternator) is on the left end, and the exciter is the smaller unit on the extreme left end (the exciter

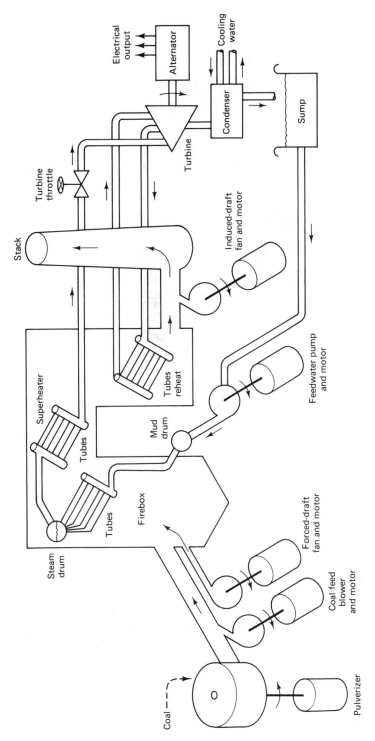

Fig. 7-2 Functional diagram of a coal-fired electric generating plant (precipitator to clean stack gases not shown).

121

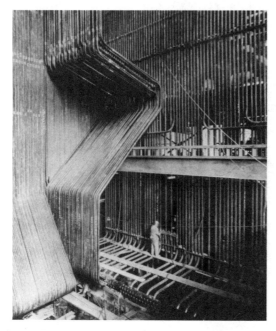

Fig. 7-3 Interior of a large steam generator. (Courtesy of AEP/Appalachian Power.)

is a generator for supplying dc current to the alternator field, to be discussed later).

Controls become quite complex for a plant of this size; the proper turbine operating conditions must be maintained and furnace operating temperatures and firebox pressures must be controlled to avoid damage. These plants operate very close to the limits of high temperature and high pressure; fast and accurate automatic closed-loop controls are absolutely required because it is impossible for a human operator to operate properly all the machinery: valves, fans, and pumps. Sometimes, as described in Chapter 2, small boilers are manually controlled, but the large installations using superheat and reheat cycles could not safely be manually operated near the maximum power output possible. A typical control room is shown in Fig. 7-5.

The efficiency of these large plants (defined as the ratio of electric power output divided by the fuel energy input) is highest when operated at the maximum capacity. That means that the utility company will prefer to operate the large boiler turbine units at 100% rating or even a bit higher. Actually, "nameplate rating" from the manufacturer is an estimate, and the owner (the utility) will want to determine what the maximum output can be. The plant engineer will experiment with slight

Fig. 7-4 Turbine-generator. (Courtesy of AEP/Appalachian Power.)

variations in steam conditions (superheat and reheat) and fuel/air ratios to reach the highest plant output and efficiency.

The heat content of the fuel used will vary (coal, oil, or gas), so the plant operator must be alert and adjust the combustion air flow control to maintain proper firebox temperature—if a particular load of coal burns hotter than usual, for example. Additional machinery is required to remove solid matter (ash) from the stack gases. In a large plant, up to 70 tons of ash is collected per hour.

Fig. 7-5 Control room of a power plant. (Courtesy of Honeywell Industrial Controls Division.)

OPERATING DATA ON TYPICAL POWER PLANTS

Units 1 and 2
  Each electric generator: 800,000 kilowatts
  Each turbine generator: 178 feet long
  Steam generator: 20 stories high
    Pressure: 3500 psi
    Steam temperature: 1000°F, reheated twice
    Steam flow: 5.3 million pounds per hour to each turbine

Auxiliaries
  Fans: each unit (2) 9000 hp and (2) 3500 hp for furnace draft
  Coal pulverizers: each unit has six pulverizers, which can each grind
    120,000 pounds per hour
  Feedwater pump: each unit has a feedwater pump which can deliver
    12,500 gallons of water per minute to the steam generator

Unit 3
  Electric generator: 1,300,000 kilowatts
  Steam generator: 25 stories high
  Steam pressure: 3675 psi
  Steam temperature: 1000°F, reheated once
  Steam flow: 9,775,000 pounds per hour

Auxiliaries
  Fans: (3) 9000 hp and (3) 5000 hp for furnace draft
  Coal pulverizers: 12 units, which can each grind 120,000 pounds per hour

Coal burned per day (all three units)
  (345) 80-ton railroad cars (27,600 tons)

Fig. 7-6 Selected data: Amos plant of Appalachian Power Co.

Some of the operating data for a large utility electrical generating plant are shown in Fig. 7-6; the John Amos plant includes three boiler-turbine-generator units. Units 1 and 2 are large (800,000 kW each), but unit 3 is even larger: 1,300,000 kW. Note the steam conditions: 3675 psi and a flow of 9,775,000 lb/hr. Even more impressive is the size of the fans required to supply combustion air: three 9000-hp and three 5000-hp units. Coal is ground into face-powder consistency by pulverizers and blown

One steam generator unit:
  Electricity produced: 1,300,000 kW
  Steam generator:
    Pressure 3845 psi
    Temperature 1000°F
    Flow 9,775,000 pounds per hour

Auxiliaries:
  (3) 13,500-hp forced-draft fans
  (3) 5000-hp primary air fans
  (3) 3000-hp gas recirculating fans

Coal burned per day
  (8) river barges, carrying 1500 tons each
              or
  120 to 140 railroad cars carrying 80 tons each

Stack for cleaning system
  Electrostatic precipitators remove 70 tons of fly ash per hour

Fig. 7-7 Selected data: New Haven plant of Appalachian Power Co.

into the furnace, where it burns very rapidly; these pulverizers can grind 120,000 lb/hr each, and there are 12 of these in the John Amos plant.

Another large unit is operated by the Appalachian Power Company near New Haven, West Virginia, as summarized in Fig. 7-7. This plant consists of a single boiler-turbine-generator unit which generates 1,300,000 kW. The coal consumption per day is 120 to 140 railroad cars carrying 80 tons each. The ash removed from the stack gases is phenomenal: 70 tons of ash per hour.

CONTROLS

A primary concern is to maintain enough water flow through the "water walls" and furnace tubes to avoid hot spots and possible melting. The furnace gases may reach 3500°F and can cause considerable damage if adequate water flow is not provided. Safety systems must cut off fuel flow quickly if feedwater flow drops below a specified minimum.

The most important control loops are as follows (see Fig. 7-2 for reference).

1. Feedwater flow is controlled by steam flow and corrected for steam drum water level. This loop is exactly as described for a boiler: the control sets feedwater flow equal to steam flow (in pounds per hour) and then trims feedwater flow, up or down, if drum water level is not correct.
2. Fuel flow is adjusted according to desired steam temperature and pressure.
3. Fuel/air ratio: Airflow is adjusted to obtain the specified ratio, so that complete combustion of fuel is obtained—but not excessive air, because that reduces furnace temperatures and thus efficiency.
4. Steam superheat temperature is controlled by spraying water into the steam pipe, leaving the superheat furnace section. If steam temperature is too high, water will cool the steam.
5. Reheat steam temperature may be controlled by valves which allow a portion of the steam flow to bypass the reheat sections.
6. The turbine throttle is adjusted to obtain the desired electric power output. The utility system dispatcher will specify how many megawatts each generating station must supply, so that the plant operator can set a control loop to maintain that output.
7. Furnace draft is supplied by both forced-draft and induced-draft fans, to maintain proper air pressure and complete combustion in the firebox. The forced-draft fan blows into the lower portion of the

firebox and the induced-draft fan blows essentially up the stack, causing low pressure to pull gases through the furnace.

The series of control loops described above maintain desired fur- nace and turbine operating conditions for steady operation, but what happens when the electrical load on the generator changes suddenly due to a normal electrical load change or maybe a transmission-line fault? This is a serious question because excessive furnace temperatures, fur- nace pressures, and steam pressure conditions may occur very quickly before closed-loop controls can reestablish proper operating conditions after the disturbance. This is the subject of transients, and involves study of every component.

CONTROL REQUIREMENTS

How quickly does internal firebox temperature rise when steam flow changes? How quickly can fuel flow be reduced? How quickly can all control loops operate to restore proper operating conditions? Studies of these problems have led to improved control schemes. The method de- scribed so far has been "boiler-following"; that is, the turbine throttle is adjusted, causing steam flow to change, which causes steam pressure and temperature to change. The boiler control loops then react to make corrections to fuel flow, airflow, feedwater flow, and superheat condi- tions; the boiler is trying to follow whatever happens at the throttle. This is basically a slow-responding system, because the boiler controls simply make corrections after several variables have become incorrect.

PLANT RESPONSE WITH BOILER-FOLLOWING CONTROL

Figure 7-8 indicates the type of behavior that results with boiler-follow- ing control:

1. Steam flow is rapidly reduced to a new level (2) from its original level (1) due to a turbine throttle change.
3. Steam pressure starts to rise (3); feedwater flow is reduced (4) be- cause its control loop senses the change in steam flow.
5. Steam pressure and furnace temperature reach peaks (5) and then start returning to proper levels; the steam pressure control loop causes fuel input to decrease (6) along with combustion air.
6. Fuel flow continues to decrease (6) because furnace temperature and steam pressure are still too high.

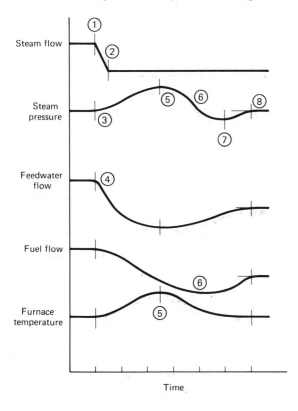

Fig. 7-8 Plant response using boiler-following control.

7. Furnace temperature reaches normal value; steam pressure may undershoot (7) before reaching normal value (8).
8. All variables reach normal values.

Note all of the delays: Fuel flow does not change until steam pressure and furnace temperature are far from normal values; eventually all of the controlled variables are correct, but the variations may be intolerable.

Considerable time is involved before all operating variables are returned to normal values. How high these variables go before being corrected is of serious concern to the plant engineer. Safety valves are always installed, so that excess steam is vented outside to the atmosphere if dangerous steam pressures occur. Naturally, it is considered poor operating practice to have the safety valves blowing—for several reasons, not the least of which is the expense of the ultrapure water that would be lost.

Improved Control

Much improved control can be obtained if the turbine change signal is sent to the boiler at the same time; in other words, change fuel flow up or down according to throttle changes *before* furnace conditions change too much. This is a form of anticipation (actually feedforward, as compared to feedback), in that the boiler conditions are adjusted upward or downward without actually waiting for boiler conditions to deteriorate to hazardous levels.

This is called a parallel scheme because power plant electrical demand signals (or steam flow signals) are sent to the turbine and boiler at the same time. Much improved control results, because the boiler is really a slowly responding mass of steel and water. By starting the change in fuel flow according to steam flow, or the turbine throttle signals, the boiler power level can be started in the correct direction before furnace conditions vary excessively.

PLANT RESPONSE WITH PARALLEL BOILER-TURBINE OPERATION

Figure 7-9 indicates plant performance that results with the improved scheme. The sequence of events is as follows:

1. Steam flow is reduced (2) from its original level (1), as before, due to throttle changes.
3. Steam pressure starts to rise; feedwater flow is reduced (4), as before.
4. Fuel flow is reduced immediately, from the steam flow signal.
5. Steam pressure and furnace temperature reach a peak (5) but much lower values than before with boiler-following operation.
6. Fuel flow begins to rise (6) as steam pressure and furnace temperature approach normal levels (7).
7. Normal operating levels are reached (8).

Note that the peaks of steam pressure and furnace temperature are lower than before, and a much shorter time is required to return to normal conditions with the parallel control scheme. Different boiler manufacturers have evolved their own versions of parallel control, which differ in some details. The overall idea is the same: Push the boiler operating level in the same direction as the generator electrical load, *at*

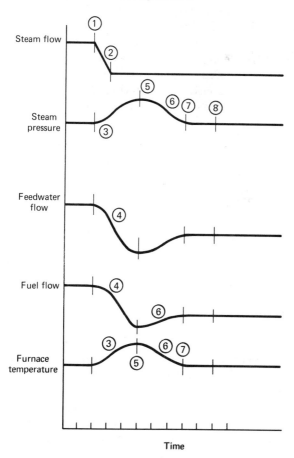

Fig. 7-9 Plant response using parallel control.

the same time, without waiting for boiler operating conditions to deteriorate.

TURBINE-FOLLOWING CONTROL

There is another variety of feedforward, called turbine-following, in which the boiler control system receives a power-level change signal before the turbine, and then the turbine is forced to take all of the steam generated, to maintain steam pressure at proper values. This method would be better for the boiler, but the utility system dispatcher probably would not like it, because the turbine-generator electrical output changes would be slow—simply because the generator output would not change until the boiler responded and increased steam flow. Each utility has its own favored type of operation, however.

Transient response (how fast the plant responds to changes) is very important because realistically the electrical load is never exactly constant and all of the control loops are always working, trying to reach zero error. The boiler is the slowest component—therefore, improved, advanced control methods are important in making the overall plant response to electrical load variations as quick as possible.

ELECTRICAL CONSIDERATIONS

The electrical output from these large generating stations is carried at very high voltages, from 230,000 to 765,000 V. Figure 7-10 shows a typical tower carrying one three-phase line; each line is actually two conductors in parallel, carrying more than 2000 A each. The two uninsulated wires on the top of the tower are ground wires, for lightning protection.

Rotary Field and Exciter

Large alternators are usually built inside out: that is, the field is rotating and the armature is stationary, as shown in Fig. 7-11; that is done for the simple reason that field current is much smaller than the generated output current, so slip-ring problems are minimized with smaller currents. The stator, then, is the output of the generator, and requires no sliding contacts, as with slip rings. The dc power required for the field is supplied by an exciter, which is a dc generator on the same turbine shaft. This generator may be several kilowatts in size.

Fig. 7-10 High-voltage transmission line carrying three-phase power. (Courtesy of Virginia Power.)

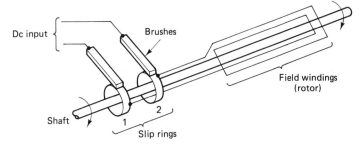

Fig. 7-11 Rotating field of a large alternator.

Each turbine-generator combination is usually wired in a delta configuration as shown in Fig. 7-12; the generated voltage may be 7600 V up to 26,000 V. The step-up transformer shown in Fig. 7-13 may raise that generated voltage to 230 kV or up to 765 kV, depending on the utility company. The circuit breaker shown is the main connection between the alternator and the utility network.

Synchronizing the Alternator with Outside Lines

An interesting operational problem arises when the boiler-turbine alternator plant is up to speed and ready to supply electric power. There is a large circuit breaker to connect the alternator electrical output to the outside lines and distribution system, but that circuit breaker cannot be closed until the plant output voltage, frequency, and phase match those of the outside utility lines. (Only one phase will be discussed; actually, generated power is usually three phase, but the principles are the same.)

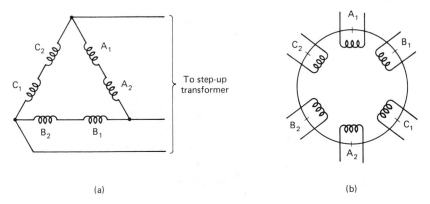

(a)

(b)

Fig. 7-12 Stator connections of a large alternator: (a) coil connections; (b) schematic of coils.

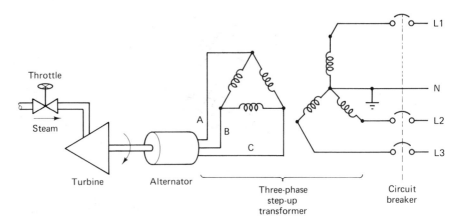

Fig. 7-13   Plant electrical output connections.

Figure 7-14 indicates the situation when the alternator output is not in phase with the outside lines. If the circuit breaker were closed with these conditions, excessive current would flow because the alternator is about 180° out of phase with the outside lines and it would be essentially a short circuit on the alternator, and could cause physical damage. The alternator speed must be carefully adjusted to obtain exactly the same frequency and same phase as those of the outside lines, as shown in Fig. 7-15, before closing the circuit breaker.

The alternator field current controls the voltage output of the alternator and the turbine throttle controls the speed, and thus frequency, of generated voltage. The plant operator adjusts field current to obtain the proper voltage, then adjusts the turbine throttle to obtain proper frequency (3600 rpm for 60 Hz on most large alternators). If the phasing is not identical to the outside line, the turbine speed is changed slightly to allow phase to change, until correct.

Phase Monitoring

How is phase monitored? A very simple scheme can be used, as well as complex electronic equipment, to indicate to the operator when conditions are correct for closing the circuit breaker. The simple scheme consists of light bulbs connected across the breaker, as indicated in Fig. 7-16. When the alternator voltage is correct, but frequency is off slightly, the lamps will flash slowly on and off. Bright lights indicate 180° phase error and dark (off) lights indicate proper phasing. The operator adjusts the turbine throttle very carefully until the lights flash on and off very slowly, then closes the circuit breaker when the lamps are off.

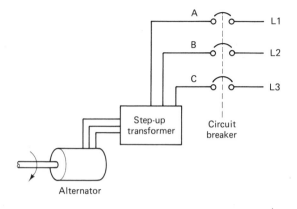

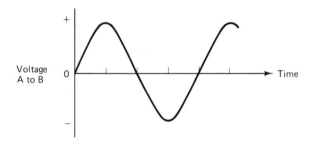

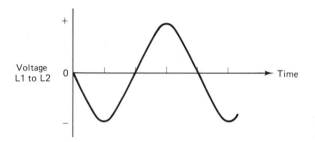

Fig. 7-14   Alternator out-of-phase condition (one phase shown).

Similarly for three-phase power: Three lamps can be used; the three lamps will flash in sequence if phasing is incorrect. When all are off, phasing is correct and the circuit breaker can be closed. Recently developed electronic systems can accomplish this procedure and close the circuit breaker automatically at the proper time, but some engineering departments may still prefer the manual method because of the ever-present possibility of system failure and improper operation of the electronic synchronizer, resulting in damage.

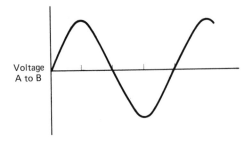

Voltage
A to B

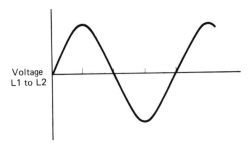

Voltage
L1 to L2

Fig. 7-15 Proper phase of voltage relations before closing circuit breaker (one phase shown).

Load Control

Once the circuit breaker is closed the alternator will run at exactly 3600 rpm and cannot vary because of the utility network to which it is connected (60 Hz, generally, in the United States). If the turbine throttle is closed, the alternator will continue to operate as a motor, driven by the utility power lines. If the throttle is opened further, to supply more steam to the turbine, the turbine will try to speed up but cannot, because it is such a small fraction of the utility system power capacity, which runs at 60 Hz.

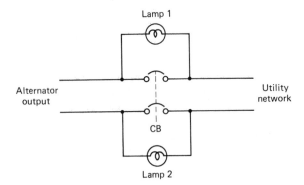

Fig. 7-16 Synchronizing lamps for a single-phase system.

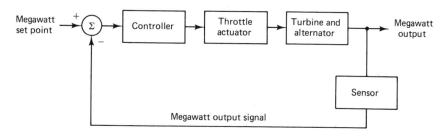

Fig. 7-17   Electrical power output control loop.

What does happen, however, is that the 60 Hz generated will be phase advanced, and the alternator will supply power to the utility network. Thus the turbine throttle becomes the load control (rather than a speed control) and determines how much electrical power is delivered to the outside network. The throttle causes the turbine to deliver more torque, which tries to raise the alternator speed and frequency compared to the utility lines. The utility network is supplied by many power plants, generating many thousands of megawatts. One alternator cannot change the frequency of the network! Each alternator, however, delivers power to the network, based on how much torque is delivered by the turbine, which depends on throttle opening. The utility system dispatcher will specify to each operating plant how much power to deliver, so that individual plant operators can then adjust a set point on a closed-loop control, which controls the throttle, as indicated in Fig. 7-17.

AUXILIARY SYSTEMS

Several other operating systems are required, not directly concerned with furnace conditions and electrical output, but of vital importance. These systems use up to 10% of the power generated in the plant; some of the more important systems are as follows:

1. Water purification and makeup system
2. Hydrogen cooling system for the alternator
3. Lubrication oil pressure system
4. Stack gas cleaning
5. Soot blowers
6. Coal-handling system

A water purification system is required, even though most of the water is condensed and reused in the steam cycle (see earlier in this chapter; typically over 9 million pounds of water per hour is pumped around the boiler-turbine loop); this water must be ultrapure, free of minerals, and free of dissolved oxygen ("deaereated"). Any minerals present could deposit on the turbine blades, causing unbalance, or in the furnace tubes, causing hot spots because water could not cool some areas. In addition, the pH (acid-base balance) of the water must be controlled, so that corrosion problems are not encouraged.

In most large installations the hydrogen cooling system is used for cooling the inside of the electrical generator (alternator). The clearances in these machines are very small, so air in that space creates friction, called "windage"; hydrogen is a very light gas, so friction due to windage can be greatly reduced by using hydrogen as a coolant. A lubrication oil pressure system is very important for the rotor bearings of the turbine-generator; without oil pressure, damage will occur very quickly due to the weight of the rotor (many tons).

Stack gas cleaning systems are required to catch the solid matter that remains after coal is burned. As indicated earlier in this chapter, up to 70 tons of ash per hour can be removed from the stack gases with electrostatic precipitators. Disposal of this waste material is a continuing problem; some of it can be used for filler in concrete, but mostly it is dumped for land fill where convenient. Many utility systems have modified their boilers so that either oil or gas can also be used, depending on prices, but coal is usually the preferred fuel, particularly on the east coast of the United States.

Soot blowers are required to clean off soot and ash from the furnace tubes, while operating; considerable material will stick to the tubes and must be removed periodically. Soot blowers are usually long tubes, 30 to 40 ft or longer, with drilled holes along one side; these tubes are then

Fig. 7-18 Chesterfield power station showing coal stock pile. (Courtesy of Virginia Power.)

Fig. 7-19 Rotary railroad car dumper; incoming coal is handled by a dumper that flips up to 100-ton cars to empty them at Virginia Power's Chesterfield power station. (Courtesy of Virginia Power.)

pushed into and across the boiler tube spaces so that high-pressure steam can clean the tubes. Special machinery has been developed to push and rotate these blower tubes to accomplish the job with a minimum loss of steam (which must be replaced by the water purification system).

The problems of handling coal are not trivial; coal must be unloaded from railroad cars, or river barges, moved to storage, and then moved onto conveyors which carry it to the pulverizers. An aerial view of a large utility plant is shown in Fig. 7-18. Railroad cars are unloaded on the lower left; coal is moved on conveyors to the top of the storage piles. Bulldozers may be seen on the piles, mixing various grades of coal. Other conveyors carry the coal to pulverizers in the building. For handling rail cars, this plant has a rotary dumper, which simply turns the railroad car upside down to empty it: 60 tons, or more, of coal (see Fig. 7-19). Other plants dump the cars from the bottom, onto an underground conveyor. Specialized machinery such as that shown in Fig. 7-20 is developed for each plant, to move up to 27,000 tons per day from storage to the pulverizers.

Fig. 7-20 Chesterfield power station: coal-handling machinery. (Courtesy of Virginia Power.)

COMMUNICATIONS

Why discuss communications in a book on control systems? The reason is that data and control signals are carried over communication links. The utility system dispatcher sends out demand signals to several electrical generating plants and receives signals (data) indicating power flow and voltage at many points in the network. An electrical utility system may be spread out over several states, with a central control room. The system is thus many spreadout closed loops. Figure 7-21 illustrates some of the signals required for measurement and control of a portion of a utility system.

The utility system dispatcher must coordinate all of the generating capacity with the electrical loads, and maintain proper voltage and frequency for all users, despite load variations. Communications are *vital*, and accuracy and reliability of data are most important, with voice backup as a last resort. Figure 7-22 shows a utility control room, where the system dispatcher monitors and controls power flow. Note the large system map on the back wall, showing all generating stations, transmission lines, and substations.

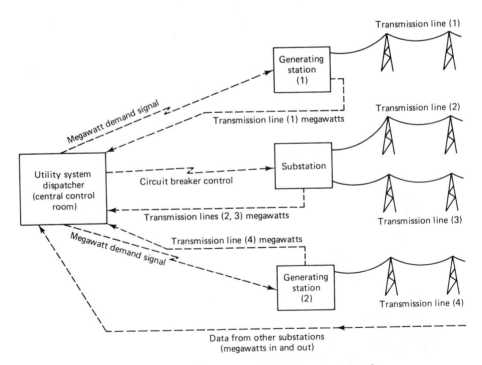

Fig. 7-21   Electric utility system: Data and control signals.

Fig. 7-22 Power coordinators work at CRT consoles in the American Electric Power Service Corporation's new System Control Center to meet the electric energy needs of the seven-state AEP System's 2.5 million customers. In the background is the Center's map board, a schematic display of AEP's generation and high-voltage transmission system. (Courtesy of AEP/Appalachian Power.)

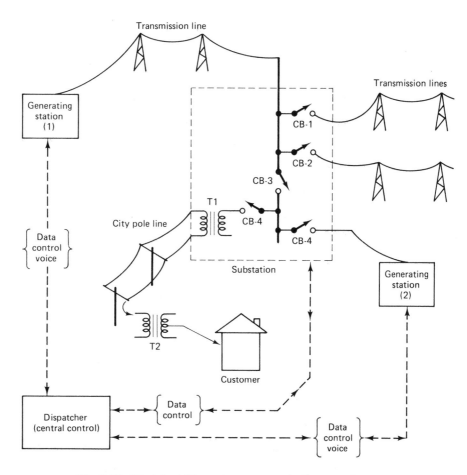

Fig. 7-23 Electric utility system: communications requirements.

Fig. 7-24 John Beachman, of Columbus, manager of the Power Control Section of the American Electric Power Service Corporation's System Operation Department, observes operation of the System Control Center's map board, a schematic display of the AEP system's seven-state generation and transmission network. (Courtesy of AEP/Appalachian Power.)

Some of the dispatcher functions are being done by computer; one reason for this is the speed required to react when a failure occurs. If a generating station fails, the megawatt load must be taken up by the remaining stations or power must be purchased from adjacent utility systems; if a transformer fails, power must be rerouted by other lines to the affected area. These decisions require quick thinking, otherwise, the remaining generating stations may be overloaded and trip out, causing a "domino effect."

Computers are thus becoming a vital link in the closed loop, of reading data from many points of the utility system and sending out demand signals to generating stations and also demand signals to major circuit breakers to open or close, as indicated in Fig. 7-23. Electrical power flow from generator to customer is shown, with communication links carrying data, control signals, and voice. Figure 7-24 is a closeup of a system map with main transmission lines and substations shown.

COMMUNICATION REQUIREMENTS: DATA TRANSMISSION

As discussed previously, a central dispatcher may send out megawatt set points to many control loops and receive data from those loops. Modern control equipment not only controls, but collects data for transmission to a central location. The system dispatcher must know at all times the operating status of each generating plant and major substations, how much power (megawatts) is being generated (or purchased from other utility systems), voltage and frequency on the transmission line, and so on. Communication is needed for control signals and data to and from each generating station as well as voice contact.

Several types of communication systems are used:

1. Telephone lines, leased from telephone companies
2. Radio signals transmitted over the utility power lines (carrier radio systems)
3. Radio communications using VHF frequencies
4. Radio communication using microwave frequencies

Each of these has advantages and disadvantages; evaluation must be done on the basis of cost and reliability as well as purely technical specifications. Consider what signals must be carried:

1. Voice
2. Data
3. Control signals

Voice transmission is relatively simple and slow speed; data and control signals, however, may be high-speed digital signals.

Telephone Lines

Telephone lines are relatively inexpensive and maintenance costs are handled by the telephone companies. However, signal transmission may be too slow for high-speed data systems. Worse yet, reliability is a major concern; telephone lines may be damaged by storms or auto accidents, or disconnected by mistake.

Carrier Systems

Carrier systems are superior to telephone lines. Radio signals are transmitted over the 60-Hz power lines. These are low-frequency radio signals (40 to 300 kHz) coupled to the power line as if to an antenna. However, there is not much radiation of these signals. The system is supposed to be tuned as a transmission line, so that the radio signals travel along the lines and do not radiate. At each substation, tuned circuits can pick up the desired radio-frequency carrier signal from the 60-Hz power line. Power lines are relatively rugged compared to telephone lines, so reliability is better. The utility system has maintenance responsibility and under normal conditions carrier systems operate well for voice and some control functions.

However, in time of major problems such as floods and hurricanes, many power lines are damaged and communications may be lost or be undependable. Noise may be a problem and interfere with data communication accuracy.

VHF Radio Systems

VHF radio transmission has gained in popularity for voice communication between control room, maintenance trucks, and individual personnel with walkie-talkies. Battery-backed power supplies are easy to arrange, so reliability is excellent. For long-range communications, high-power repeaters are possible, so that simple walkie-talkies can talk to a distant dispatcher by way of the repeater. Thus all maintenance personnel can keep in touch with headquarters.

Microwave Radio Links

Data and control signals can be carried on the systems noted above, but recently, microwave radio systems have begun to show the following advantages:

1. Lower-power transmitters are required, due to the highly directional microwave antennas used.
2. Greater bandwidth, which means that more data can be carried, at high speeds, than on carrier systems and VHF radio.
3. Better security is possible because the small beamwidth of the directional antennas results in a lower possibility for spurious pickup, which causes false control signals.
4. Reliability is also improved because the low-power microwave radio equipment runs cooler than high-power VHF equipment, and less battery capacity is required for backup.

SUMMARY

Large electrical generating stations require complex control systems to coordinate electrical output with furnace conditions. Many closed-loop controls are required to control the tremendous energy and large machinery involved. Good original design and good maintenance are vital to keep such a complex plant operating. Safety systems are required to prevent plant damage if excessive pressure or temperature occurs for any reason. A reliable communications network is required, to close the loop between the dispatcher load demands and actual load conditions.

Postscript

Many larger plants are now being built in a slightly different manner: The steam drum is omitted, creating what is termed a "once-through steam generator." There is no drum level to serve as a control input; only steam pressure, steam flow, and temperature are available. Control is therefore somewhat more difficult. Response of this type of boiler (now called a steam generator) is faster because there is no steam drum with its large mass of water and steel. The control responses are similar to those discussed in this chapter, but faster.

STUDY QUESTIONS

1. Describe the water and steam flow cycle.
2. List several auxiliary systems required.
3. Explain the control philosophies of boiler following and turbine following.
4. Explain how megawatt plant output is controlled by the turbine throttle after a generator is synchronized with outside lines.
5. Describe a simple synchronizing system.
6. Describe the communications requirements of an electric utility system.
7. Compare the advantages and disadvantages of communications using wire lines, carrier, VHF radio, and microwave radio.

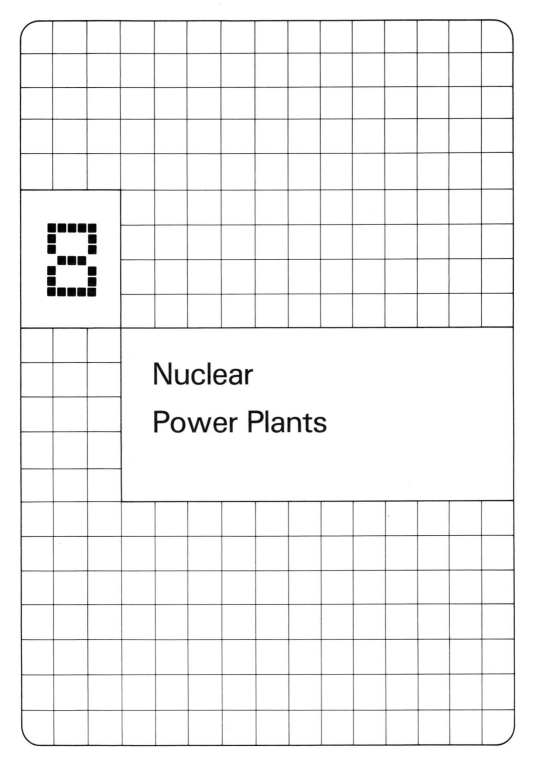

Nuclear
Power Plants

OBJECTIVES

To describe typical construction of pressurized water reactor nuclear power plants

To outline some operational characteristics and problems

To outline safety system requirements

To describe some auxiliary systems required

Nuclear electric power plants are similar to fossil-fueled plants (coal, oil, and gas fired) described in Chapter 7, in that steam is generated to operate a turbine which drives an electric generator. The heat source is a nuclear core instead of a firebox, and steam pressures and temperatures are considerably lower than in fossil-fueled plants. In this chapter we discuss pressurized water reactors (PWR), which are widely used.

CONSTRUCTION

A general sketch of a PWR plant is shown in Fig. 8-1. The nuclear core heats the water flowing in the primary loop; this water flows through one side of a heat exchanger, which generates steam on the other side (secondary). Feedwater is pumped into the secondary side, where it is heated and turned to steam. The steam flows through the turbine and is condensed and sent to the sump, to be pumped around again.

The primary loop is typically operated at about 500°F and at high pressure to keep the water from boiling (typically over 1500 psi). The secondary loop typically operates at about 400 psi, so that water boils in the heat exchanger, producing steam. This is low-pressure saturated steam, with no superheat, so the turbine efficiency is considerably lower than the efficiency obtained in the fossil-fueled plants described in Chapter 7. Generally, the higher the steam pressure and temperature, the higher the turbine efficiency. It is really ironic that the nuclear industry was forced to return to old technology to build lower-pressure turbines when the latest technology indicated that high-pressure operation is superior. However, the primary loop pressurization is a major problem and 1500 psi is difficult enough (requiring thick piping and difficult welding). Higher pressure would allow higher steam pressures on the secondary side, but there is an economic and technical compromise between expense and difficulty of fabricating a high-pressure primary loop and the advantages of higher steam pressures.

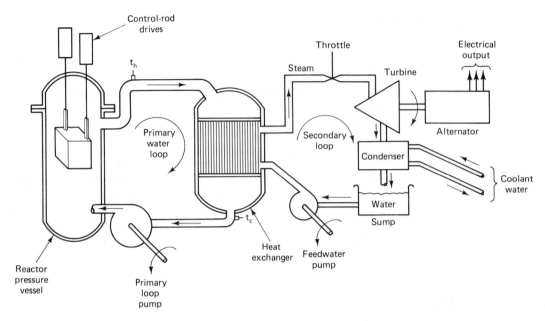

Fig. 8-1 PWR nuclear power plant.

Note in Fig. 8-1 that primary water does not mix with the second-ary side water; the primary water flows through the inside of hundreds of small tubes in the heat exchanger, whereas the secondary water and steam are on the outside of the tubes. Of course, leaks do occur, resulting in slightly radioactive steam. Normally, all radioactivity is confined to the primary side, and the steam is not radioactive.

Nuclear Core

A typical nuclear core is fabricated as indicated in Fig. 8-2; uranium oxide pellets (approximately $\frac{1}{2}$ in. in diameter and $\frac{1}{2}$ in. long) are sealed in long metal tubes. Many of these tubes are assembled into a fuel ele-ment, approximately 6 in. square and 6 ft long. These fuel elements are then mounted in the core assembly, which is a cube approximately 6 ft on each side.

A nuclear core is controlled by control rods, as indicated in Fig. 8-3; two control rods are shown, typical of perhaps 9 to 15. These are cross-shaped, to slide down between the fuel elements. As the rods are with-drawn, the nuclear fission process increases, generating heat. Power level is thus controlled by the positions of these control rods.

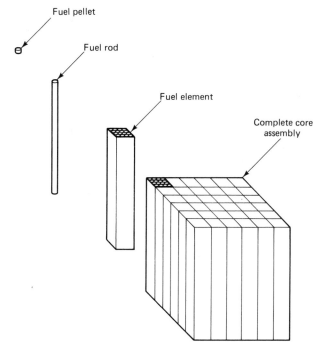

Fuel pellet

Fuel rod

Fuel element

Complete core assembly

Fig. 8-2 Nuclear core assembly.

Control-rod drive mechanisms are typically electrical drives, moved according to signals from a closed-loop control system, which compares actual core power level to desired power level and moves control rods accordingly. Most of the control rods are controlled manually, with one or more reserved for automatic closed-loop control.

Normal operation of control rods is at a relatively slow speed, say 12 in./min; emergency conditions such as over-temperature might require fast movement, so control rods must have a high-speed mode, called "scram," which drives all the rods into the core at maximum speed to reduce power quickly to a minimum. That is a function of the safety system, discussed later.

Reactor Pressure Vessel

The nuclear core is mounted inside a pressure vessel as indicated in Fig. 8-1; this is a steel "bottle," approximately 12 ft or more in diameter, made of steel 6 in. or more thick. These vessels are fabricated from flat steel plate rolled into the required cylindrical shape and welded at the seam. The top must be removable, so the dome-shaped top is bolted to

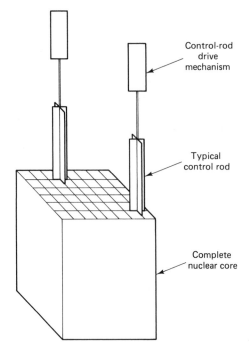

Control-rod
drive
mechanism

Typical
control rod

Complete
nuclear core

Fig. 8-3 Nuclear core control.

the lower portion with bolts that may be 4 in. in diameter. In addition, the joint is "seal welded" to ensure against leaks. (When the top must be removed, that weld is cut out.)

Control-Rod Drive-Line Seals

Additional difficulties are caused by the requirement of control rod drives, which must operate up and down through holes in the top of the reactor pressure vessel. These are sliding penetrations, which naturally could leak water due to the very high pressure primary water. A clever technique is used to handle this problem, as indicated in Fig. 8-4.

The seal consists of several stages, each with seal rings similar to those on automobile engine pistons; high-pressure pure water is pumped into the spaces between the rings. This high-pressure water (seal water) then flows both ways: into the reactor vessel, and out to the atmosphere. The idea is to control the leak, which cannot be prevented, so that the only water that leaks is pure water. Thus no radioactive water leaks out of the reactor pressure vessel.

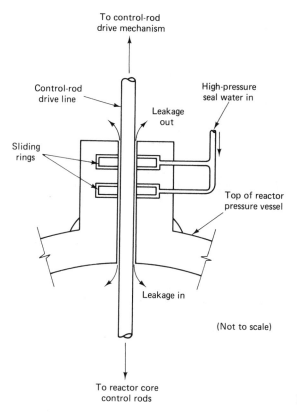

To control-rod
drive mechanism

Control-rod
drive line

High-pressure
seal water in

Leakage
out

Sliding
rings

Top of reactor
pressure vessel

Leakage in

(Not to scale)

To reactor core
control rods

Fig. 8-4 Control-rod drive-line seals.

Containment

All of the nuclear plant components, from reactor vessel to heat ex-
changers, are enclosed in a containment building, which is typically
reinforced concrete, designed to keep any possible leakage of radioactive
material from escaping into the outside atmosphere. The turbine and
generator are outside the containment.

TYPICAL CONTROL LOOPS

There are many variables to control; the control room photograph in Fig.
8-5 indicates the complexity.

1. *Reactor power level.* The power level is controlled according to
the electric power required (see Fig. 8-6); actual reactor power level is

Fig. 8-5 Control room of the North Anna Nuclear Power Station near Richmond, Virginia. (Courtesy of Virginia Power.)

sensed by ionization chambers near the core. This signal is compared to the desired power level, creating an error signal that the controller uses to determine control rod movement.

   2. *Feedwater flow.* Feedwater flow is determined by steam flow, exactly as in the fossil-fueled steam boiler, with a trimming (correcting) action for water level on the secondary side of the heat exchanger.

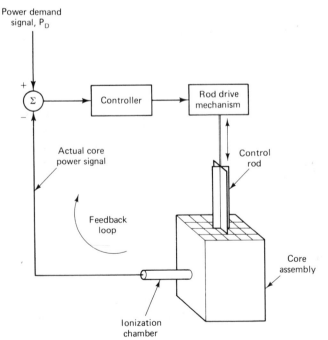

Fig. 8-6 Reactor power control loop.

3. *Primary loop pressure.* The pressure must be kept high enough so that there is no boiling in the primary loop; this is accomplished by electric heaters in the pressurizer, and by the charging pumps which inject water into the primary loop.

CONTROL PROBLEMS

One of the main difficulties in controlling a nuclear plant of this type is due to the flow time from the nuclear core to the heat exchanger. Any changes in core power produce water temperature changes, which reach the heat exchanger several seconds later. Similarly, steam flow changes produce changes in the primary water temperature, which reaches the core several seconds later. This is a dead-time problem and can result in oscillation of reactor power and even instability if the controls are not properly designed and carefully adjusted.

The usual mode of operation is to maintain the primary loop average temperature constant. Temperature measuring elements are located as indicated in Fig. 8-1. The primary loop average temperature is computed as follows:

$$T_{\text{avg}} = \frac{T_h + T_c}{2}$$

and the temperature error is computed as follows:

$$T_\epsilon = T_{\text{set}} - T_{\text{avg}}$$

A typical reactor loop demand signal consists of the following items:

$$P_D = (1) + (2) + (3) \quad \text{(control equation)}$$

where (1) is a steam flow signal, which is a good indicator of plant power output and is a relatively fast signal, from venturi flow sensors

(2) is a primary loop average temperature error signal, used to correct reactor power output so as to bring primary loop average temperature back to desired set point; it acts as a trimming action

(3) acts as a reset term as described in Chapter 3, continuing to rise as long as there is a temperature error, causing the power demand term to rise; it is a long-term correction, acting slowly to change reactor power demand if the first two terms do not result in correct average temperature

The control equation thus has three terms: the first produces about 90% of the demand signal, the second about 8%, and the third 2%. It might seem that the first term should be 100%. This would be true, except for the problems with dead time; 100% for the first term is likely to produce overshoot and oscillation in the system. With perhaps 90% in the first term, the system does not overshoot and time is allowed for the primary loop temperature to settle out before the temperature corrections are brought in by the second and third terms.

The first term causes immediate changes in reactor power demand; the second term brings in power demand changes when the primary loop average temperature drifts from the set point, and the third term brings in slow, long-term corrections to power demand in order to reach zero error in primary-loop average temperature. Note that this is a parallel control scheme (see Chapter 7). The power source (core) and the load (turbine) are driven together; the demand signal is sent to both at the same time.

There is always control system difficulty caused by the use of heavy thermometer wells on the temperature sensors in the primary loop. The high-pressure primary loop requires thick wells, causing slow response of the temperature sensors. Therefore, primary loop temperature cannot usually be used directly for reactor power control because the indicating signals are too slow, but only for slow trimming of the average primary loop temperature (the second and third terms of the control equation). Reactor power is proportional to temperature rise through the core, but this measurement is too slow to be useful.

All of these control problems are usually studied through the use of computer simulation. The effects of flow time, thermometer time constants, control-rod speeds, and other variables can conveniently be investigated and specifications written for all of the control equipment. In addition, safety studies can be done to determine the effects of too-high rod speeds, more than one rod moving, and other possibilities.

Operator training is usually done with a simulator using a full-size operating control panel; all standard operating procedures and emergency procedures can be learned before the actual plant construction is complete.

SAFETY SYSTEMS

Safety systems are required to monitor several important plant variables, such as reactor power level and primary loop pressure, and cause an automatic plant shutdown (scram) if hazardous conditions are approached. For example, the nuclear core power level is typically sensed by three channels of measurement equipment; a scram is initiated if two out of three of these channels indicates excessive power. The two-out-of-three idea reduces the likelihood of false plant shutdowns due to faulty signals and also increases the reliability; one channel can fail, and the other two can initiate shutdown action if required. Safety systems are always battery backed, so that shutdown can be initiated even in case of major trouble when electric power is lost.

AUXILIARY SYSTEMS

Several other systems are required to maintain proper operating conditions of the primary and secondary loops. The following systems illustrate the more important ones.

Primary Loop Pressurization

The primary loop pressure must be maintained at proper values at all times to avoid boiling of the water, which could allow overheating in the nuclear core and possibly cause damage. Excessive pressure must also be avoided, so the pressurizer is always provided with a relief valve, which operates similarly to the steam safety valve on a steam boiler (Chapter 7). The pressurizer operates with an air space above the water level; electric heaters in the pressurizer are used to raise primary loop pressure to desired levels.

Charging Pump/Boron Injection

The charging pump pumps water into the primary loop, in order to maintain the proper water level in the pressurizer. Boron injection is a safety backup system: In case the control rods cannot reduce core power in an emergency, boron is pumped into the primary loop to reduce core power.

Seal Water for Control-Rod Drive Lines

High-pressure water must be provided to the control-rod drive-line seals (discussed previously). This water must be at higher pressure than that of the primary loop, so that the controlled leakage is into the primary loop.

SUMMARY

Nuclear power will probably be a very important source of energy for the future, as supplies of coal, gas, and oil are not endless. A typical nuclear electric generating plant can supply energy for about three years before refueling; compare this to a large coal-fired plant, which burns 60 tons of coal in 10 minutes. Safety is always important, and reliable equipment with battery backup is vital. Separate sensors and electronic equipment are always installed, to provide protection if normal control channels fail.

Besides PWR reactors, other types of nuclear reactors are in use around the world; these include boiling-water reactors (BWRs) and gas-cooled reactors. BWR reactors generate steam directly in the nuclear core and do not have a secondary loop; The gas-cooled type uses helium or carbon dioxide as the heat-carrying substance. Each type of plant has advantages and disadvantages; the reader is referred to the Bibliography for more details.

STUDY QUESTIONS

1. Describe the construction of a typical nuclear core.
2. List the major components of the primary and secondary loops.
3. Why is a PWR plant limited to low steam pressure?
4. Explain several control difficulties of a PWR plant.
5. Describe the function of each term of the "control equation."
6. What are advantages and disadvantages of a BWR plant?

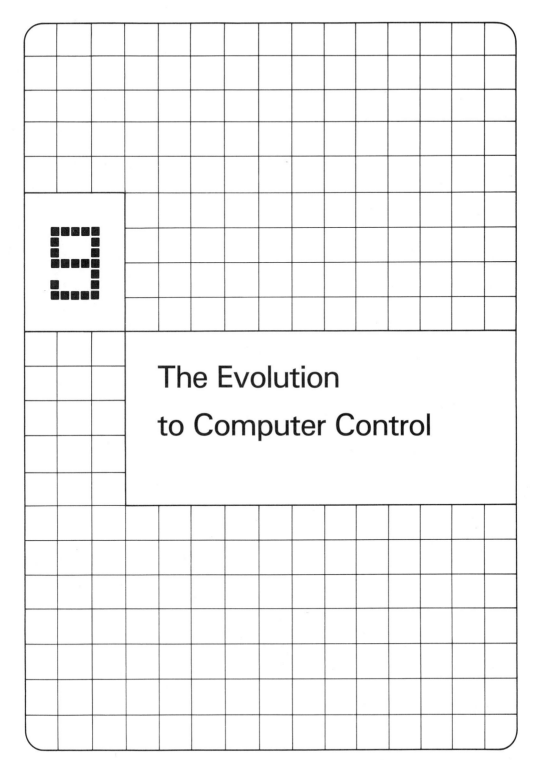

9

The Evolution
to Computer Control

OBJECTIVES

To compare analog and digital control

To explain central control versus distributed control

To illustrate typical single-loop controllers

To discuss modern loop controllers

Control systems have changed gradually from mechanical (pneumatic and hydraulic) to all-electronic controller equipment, including computers. Even the single-loop controller typically includes a version of computer called a microprocessor. Central data collection is generally included in modern distributed control systems to provide immediate information to management.

For many years most control systems used analog signals: primarily pneumatic, along with some electrical and hydraulic systems. Many of these systems are still operating; signals are transmitted as air pressures or electrical voltages. For example, a 3- to 15-psi air signal could represent a temperature range of 100 to 800° F. Electrical systems might use as current signal, 4 to 20 mA, or a voltage signal 0 to 10 V, to represent a steam pressure range of 200 to 1000 psi.

The error detector and controller circuits were originally all mechanical, with air-pressure signals; later, electronic equipment used magnetic amplifiers, then tubes, and then transistors before integrated-circuit electronic components were available.

PNEUMATIC CONTROLS

Control panels using pneumatics are a maze of small copper piping carrying the many sensor signals in, and actuator signals out. These panels are usually large, often occupying a whole wall of a control room because of the large size of the components and hundreds of copper pipes. Usually, the controller function is combined with a recorder, so that a paper record is continuously made of the variable being controlled. These are frequently round charts, to show one 24-hour time period on one chart (round charts are considered easier to file than are strip charts). The controller/recorder might require 12 by 15 in. of panel space. Later models of these controllers are made in smaller sizes, perhaps 8 by 8 in., using strip-chart paper about 5 in. wide. It is not unusual to see 30 to 50 or more of these recorder/controllers on a control panel for an oil refinery or coal-fired electric power plant, because of the many individual control loops.

Central Control Room

Note that the discussion above implies that there is a central control room to which all measurement signals come and from which all actuator signals originate. Central control is advantageous, in that all controller gain, reset, and rate adjustments are accessible in one place. Also, control equipment will be more reliable when installed in a clean environment, such as the control room, where temperatures are not extreme, and dirt and moisture are not a problem.

Some installations still have controllers located near the process (local control), which may be 100 ft or more from the control room (with only a set-point command and readout indicator in the control room). If a controller must be adjusted, a technician must be sent to make the adjustment, with communications by telephone or radio. This is an awkward procedure, to say the least.

But there are two reasons for locating the controller near the process being controlled rather than in the control room:

1. *Speed of control response.* Long pneumatic lines can cause a troublesome delay in signal transmission; 200 ft of 1/4-in. tubing might cause several seconds' delay in operating an actuator, for example. That delay might be intolerable if the process is fast responding. Good control is impossible in some cases if signal transmission time is too long.
2. *Reliability.* Long runs of copper tubing can be damaged if installed through hazardous environments (machinery, trucks, cranes), whether overhead or underground.

Thus central control evolved to a compromise: a local "blind controller" near the actual process, which simply takes orders from the control room plus a local manual controller. In case of trouble, the technician can use the local manual control to operate the process while the central control room is being repaired.

Central control may seem like a good idea, but some backup alternative is mandatory. There must be some way to maintain control of the process (which may be dangerous) in case of failure of the central system.

Advantages of Pneumatic Control

Air-operated control systems still have several advantages:

1. They are relatively simple to understand and repair. Technicians can see how they operate, with levers, springs, valves, bellows,

chains, and pressure gauges. Tools and measurement equipment required are not complex.

2. They are safe in hazardous environments. There are no electrical circuits that might cause sparks.

3. They have simple power-failure protection. Compressed air from large tanks will run the controller system for several hours in case of power failure to the air compressor. (Compare the compressed air tank to batteries required for backup of electronic equipment: Batteries require space, ventilation, and regular maintenance.)

4. Air-operated actuators and motors do not overheat when stalled. Electric motors must be very carefully selected so that they are not damaged in case of jammed linkages or stuck valves. A mechanical linkage can break, which can be required, but if the breakage causes an electric motor to stall and burn up, that may be serious, because a quick replacement may not be available.

ELECTRICAL/ELECTRONIC CONTROLS

One of the great advantages of electronic signal transmission is speed: no long signal delay as with pneumatic signals. Also, bundles of electrical wires are smaller and easier to install than is copper signal tubing. Electronic control did not quickly take over from pneumatics, however; electronic control equipment was generally higher cost, initially, at least until about 1965; maintenance can be more difficult and expensive, particularly if all maintenance personnel must be retrained; more sophisticated test equipment is required, and a spare-parts inventory must be maintained.

Industrial managers saw no particular good reason to change over to electronics if process control was adequate, as performed with pneumatic components. However, if maintenance costs can be reduced, as well as making a better product, closer to desired tolerances, then electronics has an advantage.

HYDRAULIC CONTROL

Some control systems use fluid pressure rather than air pressure for signal transmission; there are several major advantages:

1. Hydraulic control systems are much faster because fluids are essentially noncompressible. A pressure change at one end of a pipe is

transmitted very quickly. Air signals, on the other hand, are slow due to the compressibility of air.

2. Hydraulic control signals do not "bounce," due to the noncompressibility of fluids such as oil; air systems tend to overshoot due to the volume of air in signal lines and actuators.

3. Higher power can be transmitted; hydraulic pressures range from 100 to 3000 psi; thus a small hydraulic cylinder can produce very large forces. Typical applications are in aircraft control surface actuators and throttle control of large steam turbines; both of these applications require large forces, with fast, accurate, bounceless movement. A familiar application is power steering on automobiles, which is hydraulic for the same reasons: high power in a small space, with precise movement and no overshoots.

Disadvantages of Hydraulic Systems

The principal disadvantage is probably leakage, which is inevitable. Hydraulic fluids are messy; pump seals, piping, and valves will leak eventually. There may be a fire hazard with high-pressure fluids; leaks through a small hole can produce a fine mist which may be inflammable or explosive. Improved fluids can reduce this hazard. Increased maintenance may be required, simply because high-pressure systems are under more stress than is a low-pressure air system. The 15-psi air system may run "forever" without breakdown, but a 100-psi oil system may blow out seals or wear out bearings every few months.

DIGITAL SIGNAL TRANSMISSION

All of the applications discussed have been analog, whereby a temperature, pressure, or flow rate is represented by a measurement signal which varies to represent the variable. Thus a 3- to 20-psi air signal might represent 100 to 300° F, or an electrical signal of 0 to 15 V might represent the same temperature range. Digital computers have been developed and digital signal transmission has become more popular. Process signals are transmitted by binary codes, combinations of 1's and 0's. For example, 0000 to 1111 binary could represent 16 values of a variable on a four-wire signal transmission scheme; each wire would either be on or off, either 0 or 5 V, for example (see Fig. 9-1). The four wires would indicate the signal: from all off to all on, 0000 to 1111, where each lamp indicates a binary bit, either on (1) or off (0). The value of the signal would be read as follows with the usual binary weighting, 8-4-2-1:

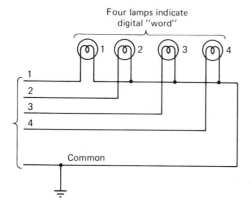

Fig. 9-1 Parallel digital signal transmission.

Lamp indication	Equivalent value
0000	0
0001	1
0010	2
1111	15

Actually, a logic 1 would be represented by any voltage from 2.4 to 5 V, and a logic 0 by any voltage 0 to 0.4 V. Thus one advantage of digital signal transmission becomes obvious: exact voltages are not as important any more, and calibration and signal drift are less troublesome. That is intentional in the design of digital systems, to make the overall system less liable to false signals.

Mention was made above of four-wire transmission, to send 4 "bits" (or pieces of information). That would be parallel signal transmission of information; another scheme is to send the 4 bits one at a time over two wires, and reassemble the 4 bits properly at the receiving end (see Fig. 9-2).

This is called serial transmission, whereby bits are sent one after the other. Of course, 4 bits can represent only a range of 16 values, so practically speaking, 8 bits, 16 bits, 32 bits, and more are used today for more data accurate transmission. Electronic circuits can be arranged to send these serial bits one at a time, and properly display them all at once at the other end. (For more depth and detail, the reader is advised to refer to some of the many books available on this topic; see the Bibliography.)

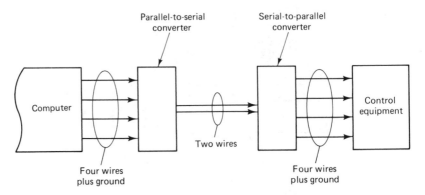

Fig. 9-2   Serial digital signal transmission.

CONTROL BY DIGITAL COMPUTER

When digital computers were first built in the late 1950s, there were predictions that central computer control would take over all process control, and indeed some installations were made in which one central computer received all sensor inputs and sent out all actuator commands. The computer acted like many separate control loops, but all in one machine. There were many problems with these first installations, among which were the following:

1. *Reliability.* When the computer broke down, nothing ran; this was not acceptable plant operating practice.
2. *Programming difficulties.* Instructions to the computer must be in the form of a program; for many years, only professional programmers could write these instructions, leaving the plant engineer and technician in the dark unless they, too, learned programming. Thus, when plant performance appeared to be improper, there was considerable difficulty in determining whether the program was faulty (software) or some piece of hardware such as interconnections, sensor, actuator, or interface had failed. Meanwhile plant production was zero.

One partial solution was to reinstall local controllers for each control loop, so that if the central computer failed, then at least the process could be safely controlled using individual loop controllers, perhaps not in an optimum manner but at least safely; also, some salable product could be manufactured while repairs were made on the computer. Plant managers do not appreciate the fact that the central computer can be a central bottleneck!

DISTRIBUTED CONTROL

As discussed above, it became obvious that centralized computer control could result in an unreliable system; reinstalling the local loop controllers provided the backup in case of central computer malfunctions; the functions of the central computer, then, are to send out set-point commands to the local loop controllers and receive data such as flow, pressure, or temperature from the local controller (see Fig. 9-3). Thus the central computer can direct the whole plant, dictating set points for each loop, and collecting data from plant operation.

The local controller illustrated in Fig. 9-3 controls flow rate; the set-point command signal normally comes from the central computer. The local controller incorporates a feedback loop and error detector, so that the valve is automatically adjusted to maintain the flow equal to the set-point command. Flow rate data are transmitted back to the computer; there is, then, a feedback to the computer so that the computer can determine if its set-point commands were followed. The local loop can also be operated locally, with a manual set-point input. This is still

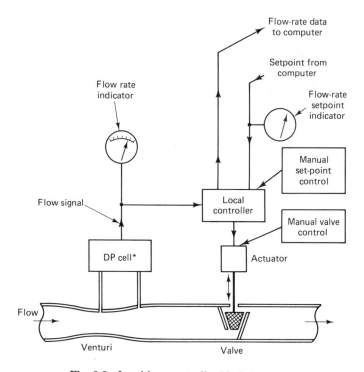

Fig. 9-3 Local loop controller block diagram.

a closed-loop flow control, and data can still be transmitted to the distant computer. If the local controller needs to be taken out of service for any reason, the valve itself can be manually controlled directly; this type of operation would be essentially open loop, with the operator watching the flow-rate indicator and moving the valve manually.

The local control loop could still be pneumatic, receiving the set-point command electrically from the computer (serial digital, probably), or the local control loop could be electronic. Note the replacement and upgrading of control equipment can be gradual; separate control loops can be rebuilt or replaced one at a time, keeping the plant running. it usually is not practical to shut down a factory for six months in order to replace and rewire all control components. It is more feasible to replace control loop sensors, controllers, and actuators one at a time. Then problems can be solved as they occur; at least you know that things were running properly before you changed one piece of hardware. If 20 controllers are replaced at once, many difficulties will probably occur, and will take time to locate and correct, because so many things were changed at one time. Of course, if a new plant is being built, there is time so that you can check out each control loop as installed, and work out most of the "bugs" before the rush is on to get everything working at once.

LOCAL CONTROL LOOP IMPROVEMENT

From pneumatic analog systems, to electronic analog systems, to a centralized computer control, loop controllers using individual small computers have evolved. Each modern loop controller contains a small computer, which performs all of the functions of proportional, reset, and rate (PID) and more; the front panel generally looks very similar to the older controller/recorder, so that operator training is not difficult. These small computers are called microprocessors, to distinguish them from large computers. The functional difference is slight, being primarily in two areas:

1. Microprocessors have a fixed small program; for example, one might perform only error calculations and proportional control; the program is stored in a read-only-memory (ROM) chip, installed by the manufacturer.

2. The microprocessor is usually "embedded" and invisible to the operator, because it is important that the front-panel controls be understandable to factory personnel. It is not important that an operator know about the internal computer operation; what is important is that the operator can easily adjust set points, gains,

reset rates, and so on. No programming knowledge is required; it is "user friendly," as the current popular phrase states.

One of the most advanced and versatile loop controllers is the Honeywell UDC-5000 (Universal Digital Controller) (see Fig. 9-4). This unit can operate as an independent loop controller or under the control of a supervisory computer through a communication interface (network operation). In network operation, the supervisory computer can specify all operating settings and request data on process performance. The basic accuracy of the controller is $\pm 0.05\%$, allowing temperature control to $\pm 1\,°\mathrm{F}$, typically.

It uses a microprocessor to accomplish many possible functions, but emphasis is on simplicity of use. In normal use, the front panel is up and only four pushbuttons are accessible (switching from manual to automatic loop control, and change of set point). The process variable is shown on the upper display and the set point on the lower display. For access to setup controls (gain, dead band, reset action, alarm set points, etc.), the front panel swings down, as shown in the figure.

There are several modes of operation, from on/off output, to PID control and proportional output. Inputs can be direct from thermocouples or resistance thermometers or millivolts or milliamperes from any sensor; in addition, math functions (add, subtract, multiply, divide, square root) are provided. Ramp/soak programming is available, so that you can

Fig. 9-4 Loop controller UDC-5000, showing setup panel. (Courtesy of Honeywell Industrial Controls Division.)

Fig. 9-5   Circuit boards of UDC-5000. (Courtesy of Honeywell Industrial Controls Division.)

program up to 10 ramp and 10 soak segments. *Ramp* means a controlled change in set point over a prescribed time, and *soak* means to hold a set point for a prescribed time. This feature would be useful in a factory process which requires, for example, a temperature change from 100 degrees to 400 degrees over a time period of 2 hours, then hold at 400 degrees for 4 hours, then cool down to room temperature in 6 hours.

A wide range of applications can be handled by the UDC-5000, simplifying the spare-parts inventory. Figure 9-5 shows the modular construction: several replaceable printed circuit boards, including the display assembly.

SMART TRANSMITTERS

The so-called "smart transmitter," also a microprocessor-controlled device, provides spectacular improvement over previous equipment. A typical application would be as a pressure sensor on a flow measuring device (venturi or orifice plate). Older designs of the electronics interface could be built only for a specified flow range (pressure drop range); other flow ranges required a different transmitter. The microprocessor-controlled transmitter allows a single device to measure flow over a 400-to-1 range. This capability has many advantages, as follows:

1. During startup of a process, flows will be small and hard to read on the normal high-range flow meter installed. Possibly another flow

meter, low range, would be required for startup measurements. The smart transmitter can be adjusted to read the low range when required.

2. Changes in process behavior (or errors in engineering design) can result in pressures different from predicted values. That might require installation of other flow transmitters. The smart transmitter can be readjusted to cover the new range.

3. Calibration is simpler. Previously, flow meters and transmitters were removed for calibration periodically; with a smart transmitter, calibration can be done in place.

4. Spare-parts inventory is simpler because one model pressure transmitter can be used for practically all process pressure measurements.

Not only are these pressure transmitters more versatile in covering a wide range of inputs; they are also more accurate. An extraordinary example is the Honeywell ST-3000 smart transmitter; specifications show a turn-down ratio of 400: 1, which means that it can be used to measure, for example, a pressure differential of 0 to 1 in., up to 1 to 400 in. of water pressure without loss of accuracy; typical accuracy is $\pm 0.1\%$ of calibrated span. The Honeywell model is four times more accurate than a standard transmitter.

Control of the ST-3000 smart transmitter is accomplished by a hand-held controller, which can remotely re-range the transmitter, modify other functions, and diagnose possible problems. No hardware changes are required; all range adjustments can be done with the controller, connected when required.

FIBER OPTICS

Fiber optics is a relatively new data transmission method whereby light signals are carried through glass or plastic fibers. The data are usually digital, so high-speed pulsed light sources are used to generate the input signals to a fiber optic cable. Receiving-end devices include photodiodes and phototransistors (see Fig. 9-6). A typical application is a data link between a central control room and the factory floor; this link can carry many control and data signals on one fiber. Of course, at each end electronic circuits are required to handle the conversion of parallel to serial, serial to parallel, and the multiplexing, whereby one fiber carries many signals.

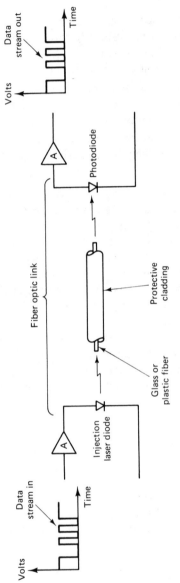

Fig. 9-6 Data transmission using fiber optics.

Advantages

Compared to data transmission using wires (shielded or coax) fiber optic systems have several tremendous advantages:

1. Immunity to interference from electromagnetic and radio-frequency sources. Fiber optic cables can be installed anywhere: next to power cables, welding equipment, and so on.
2. Ground loop problems are nonexistent.
3. Much greater bandwidth, allowing faster transmission of signals, and even multiplexing of many separate data signals on one fiber.
4. Safety; no sparks are possible as with electrical signals.

Difficulties

Losses.   A light signal is naturally reduced as it passes through the fiber. The major losses occur at the end fittings; typical cable losses will require a repeater about every 600 ft; each repeater involves a receiver, an electronic amplifier, and a transmitter.

Installation.   Installation may be difficult because proper splicing of sections of cable and attachment of end fittings is very critical; detailed procedures must be followed to avoid excessive losses of signals at these joints. Repairs to damaged cables are likewise difficult, and require special tools. Electrical cable can be tapped at several points, allowing for connection of several devices. Fiber optic cables are difficult to tap, so at present, use is limited primarily to point-to-point transmission.

Fiber optic data transmission systems have many potential advantages and will be used more as better hardware becomes available (low-loss fibers and connectors, and inexpensive fittings and tools).

SUMMARY

Process control systems have progressed from the relatively slow pneumatic equipment to central computer control and distributed control networks. Even the individual loop controllers and pressure transmitters are being improved with internal microprocessors.

Care must be taken to maintain plant reliability and avoid letting a central computer become a bottleneck. Distributed control can provide local control when the central facility fails.

Data communications methods have improved from wiring with twisted pairs, to coaxial cable, to fiber optics, which has great potential provided that the difficulties of installation and repair can be solved.

STUDY QUESTIONS

1. Explain the major advantages of distributed control.
2. Compare central control with distributed control.
3. Discuss several advantages of digital signal transmission compared to analog.
4. What is a smart transmitter?
5. What is a microprocessor?
6. How is fiber optics used to carry data communications?
7. What are the advantages of using optical data transmission?
8. What is multiplexing?

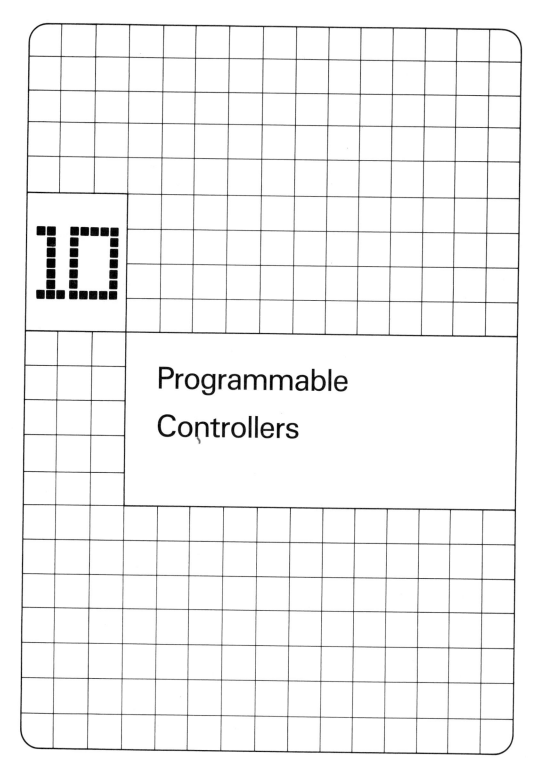

10

Programmable
Controllers

OBJECTIVES

To explain fundamentals of programmable controllers and their purpose

To illustrate ladder diagram programming

To describe some applications

To indicate some important precautions

A programmable controller is a form of computer, but the actual computer is "hidden" from the user. The user sees the programmable controller as another form of equipment which uses ladder diagrams for programming. Ladder diagrams are a convenient form of electrical schematic, familiar to engineers and technicians.

Programmable controllers were invented to replace large panels of electrical relays which are used to control sequences of machine operations. Particularly in the auto industry, the relay control systems are complex, often including hundreds of relays, contacts, pushbuttons, in-

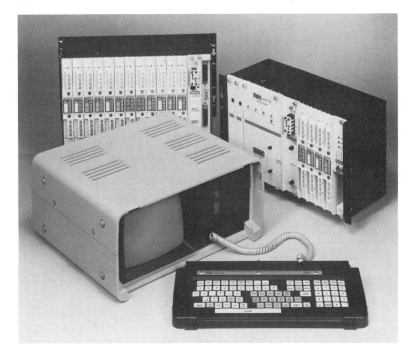

Fig. 10-1 Programmable controller system. (Courtesy of Honeywell Industrial Programmable Control Division.)

dicator lights, and limit switches. These panels require rewiring for each production model change. That rewiring can be quite time consuming.

Figure 10-1 shows a typical programmable controller system. The programming terminal is in the foreground, with the separate keyboard; the actual controller is at the right rear, with some I/O modules. Additional I/O modules are in the rack, left rear.

TYPICAL APPLICATION

A typical application could involve a production line on which engine blocks are moved on a conveyor (see Fig. 10-2). At a certain workstation, holes are to be drilled; at that position, the block is moved off the main conveyor to the drilling station and back after the required machining. A considerable number of steps are required to accomplish this sequence. Several sensor systems are required, as well as actuator devices to move the block from main to branch conveyor and from branch conveyor to machining station. A possible sequence of events (simplified) would be as follows:

1. Sensor 1: determine the presence of an engine block, and read an identifying label, probably in bar-chart format.
2. Operate branch conveyor, "forward."
3. Sensor 3: sense that the branch conveyor is clear.
4. Actuator 1: pull block from main to branch conveyor.

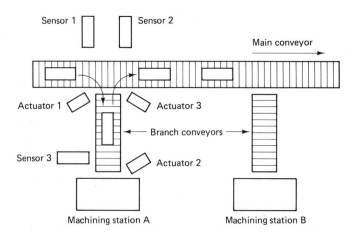

Fig. 10-2 Example of machine line requiring sequential logic and control.

5. Sensor 3: determine presence of block on branch conveyor, ready for machining.
6. Actuator 2: move block to machining station.
7. At machining station, perform the following sequence:

Clamp block to machining table
Move table to prescribed location (*x-y* coordinates)
Select drill
Operate drill
Sense depth of hole
Retract drill
Sense drill retracted properly
Move table to other prescribed hole locations and repeat as required
Unclamp block
Reverse branch conveyor
Move block to branch conveyor (actuator 2)
Sensor 3: sense block on branch conveyor
Push block onto main conveyor (actuator 3)
Sense block on main conveyor (sensor 3)
Sensor 1: check for presence of next block

All of the functions listed above, input (sensors) and output (actuators), could be controlled by a group of electrical relays wired to accomplish the proper sequence and logic, but the programmable controller can do that sequence, or accommodate changes, simply by programming. Thus hole locations, drill sizes, and drill depth changes can easily be made in the programmable controller program. If a particular engine block model does not require certain machining operations, a workstation could be bypassed. Many workstations can be involved; a typical machining line could involve 20 to 100 workstations.

Programmable controllers are a great improvement because most relays and relay wiring are replaced by computer programming stored in memory. The sequence of operations is flexible; to make changes in a machine's operation, relay sequences, pushbutton functions, or limit switch functions, no rewiring is required—only a program change.

PROGRAMMING DEVICES

Usually, a separate programming terminal is provided, such as that shown in Fig. 10-3. The ladder diagram appears on the CRT (cathode ray tube) screen and can be entered or modified using the keyboard. The symbols for switches, contacts, wires, and relays are on the keyboard, so

Fig. 10-3 Programming terminal.
(Courtesy of Honeywell Industrial
Programmable Control Division.)

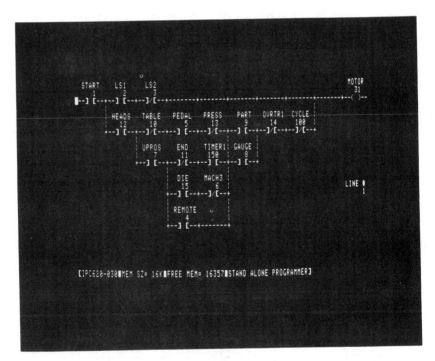

Fig. 10-4 Ladder diagram on CRT. (Courtesy of Honeywell Industrial Programmable Control Division.)

Fig. 10-5 Membrane-type keyboard. (Courtesy of Honeywell Industrial Programmable Control Division.)

each symbol is entered by typing the appropriate key. Wiring and symbols can be deleted to make wiring changes. Figure 10-4 is a photograph of an actual CRT display.

Programming is usually done "off-line," that is, not connected to the actual process. After the program is completed, it can be stored on a tape cartridge, or disk, and later actually loaded into the programmable controller computer. After that, the programming terminal is not required, and the controller will run the process without it. Thus one programming terminal is sufficient for many programmable controllers—except for the fact of reliability, for which at least two programming terminals should be available. In case of trouble, the terminal can be used to monitor the process response and observe the control functions. During startup and troubleshooting, the monitoring feature is valuable.

Note that the keyboard shown in Fig. 10-3 is built with standard open keys, similar to a typewriter; if dirt and moisture are likely to be problems, a membrane-type keyboard can be specified, as shown in Fig. 10-5. The keyswitches are operated through the flexible key covering.

PRINTER OUTPUT

Usually, a printer is part of the system, so that a permanent record can be made of the program; the ladder diagram can be printed out exactly as seen on the CRT screen, In addition, data collected in the programmable controller memory can be printed as desired.

LADDER DIAGRAM EXAMPLE

A simple ladder diagram is shown in Fig. 10-6, which represents a control scheme for a home furnace. The wall thermostat contacts (T) close when actual room temperature is less than desired. The H relay operates

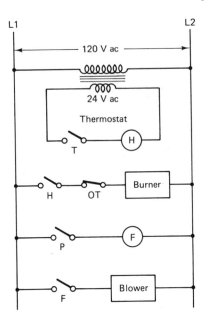

Fig. 10-6 Ladder diagram: control of home heating system.

to turn on the oil burner; when the air plenum temperature reaches a preset temperature, say 140 degrees, a thermostatic switch in the air plenum closes ("P" contact), causing the F relay to operate, which turns on the air circulating blower. A safety feature is also shown: The OT contact is an overtemperature switch in the air plenum, which opens on excessive temperature, say 180 degrees.

This circuit shows two relays, H and F; if this ladder diagram is implemented in a programmable controller, these relays do not really exist because the whole diagram is simulated by the computer. Only the five outside connections are required:

    Thermostat contacts (T) (input)

    Plenum switch contacts (P), normal (input)

    Plenum switch contacts (OT), overtemperature (input)

    Burner control (output)

    Blower control (output)

PCs are programmed with ladder diagrams, so the technician does not have to understand any computer language; the furnace controls appear to operate in a familiar manner according to the ladder diagram.

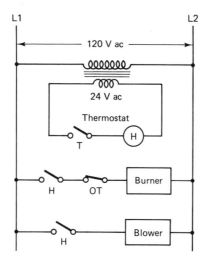

Fig. 10-7 Modified ladder diagram: home heating system.

Sequence Changes

Suppose the operating sequence is to be changed, so that the fan operates any time the burner is on; the ladder diagram would be as shown in Fig. 10-7). Note that the H relay now has two contacts, but these do not really exist except in the computer program. The ladder diagram in the programmable controller could be modified in a few minutes, with no wiring changes; only a program change is necessary. Note that the plenum temperature switch (P) is still installed and wired to the PC, but the program (ladder diagram) does not use it.

RELAY FORCING

Another useful feature of a PC is the ability to manually operate any ladder diagram contact or coil, from the programming terminal. Thus if a technician wants to test the blower motor, the F relay can be "forced" on at the controller, without operating the oil burner. This feature can be of considerable help in troubleshooting systems involving large numbers of relays; each relay can be individually "forced" to operate, and the results observed, to track down a faulty limit switch, pushbutton, solonoid, or wiring.

POWER CONTACTORS STILL REQUIRED

Of course, in most systems, some relays and contactors will still be required because the programmable controller normally cannot handle loads greater than about 10 A. Large power contactors are still required for loads such as electrical heaters and large polyphase motors.

INTERNALS

Programmable controllers are built of the same basic units as those of any computer, plus some specialized modules:

CPU (central processor unit). This is the basic computer, consisting of the CPU chip and control circuitry.

Memory. Both ROM (read-only memory) and user-programmable memory (RAM—random access memory, actually read/write memory) are provided.

I/O Input/output connections to the outside world. These may be on/off signals or analog signals, or connections to a supervisory computer.

Programming device. Sometimes a simple pushbutton device is provided; more frequently, a CRT unit is used to display the ladder diagram.

A fundamental block diagram is shown in Fig. 10-8. The CPU directs the entire operation by way of the address and data buses. The ROM contains the fixed program which converts the ladder diagram instructions into binary form for the CPU; the RAM contains the user program (ladder diagram) entered through the programming device. The battery is usually included, to avoid loss of the RAM memory content if line power fails.

The I/O modules are available in many types: ac in and out, both high and low voltage; dc in, both high and low voltage, and analog in and out (both voltage and current signals). Also shown in a remote I/O panel connected to the main processor by a coax or fiber optic link; that panel might be 500 ft from the main processor, so considerable wiring is avoided by using an I/O panel close to the process. Another coax or fiber optic channel is shown, for communication with a supervisory computer.

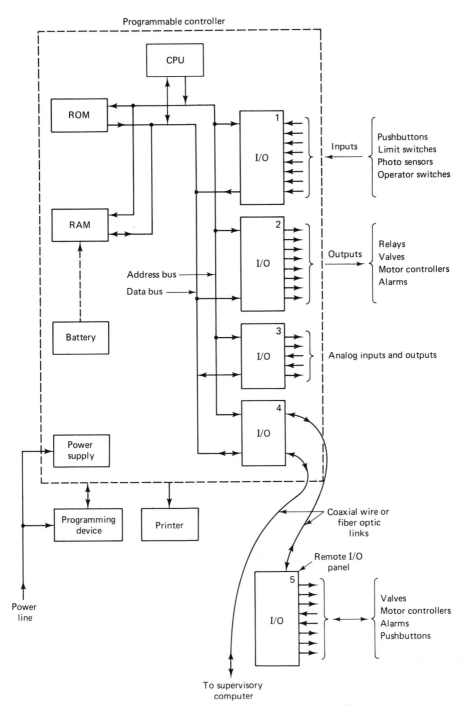

Fig. 10-8 Programmable controller block diagram.

OTHER FEATURES AVAILABLE

Besides the bare control logic for on/off relay-type functions, most programmable controllers include additional features such as the following:

Timers. These devices can be preset to accomplish time delays (delay on and delay off) and sequence of operations exactly as can be done with mechanical timers.

Counters. These devices can record, for example, how many times a drill is used, or how many times a limit switch is operated, or simply a count of product items passing a point.

PID [proportional, integral (reset), and derivative (rate), as explained in Chapter 3]. PID is available for most processors as a special module. These modules include analog-to-digital (A/D) and digital-to-analog (D/A) converters, and some allow sensors such as thermocouples or thermistors to be connected directly to the programmable controller module. The output signal (analog) is thus variable according to how much correction is required.

Math Functions. Some programmable controllers have mathematic capability (add, subtract, multiply, divide).

Graphic displays. A variety of graphic devices can display a visual model of a process showing tanks, valves, conveyors, and status of each (level, speeds, etc.).

Other programming languages. Several vendors provide other programming languages, such as BASIC or FORTRAN, which allow programming of the desired relay logic in programming statements rather than ladder diagrams. Even assembly language programming is possible on some equipment; although considerably more difficult, it can produce more efficient programs (they run faster in the computer) if the time required for execution of ladder diagram programs is excessive for your application. However, even BASIC language programming is considered more difficult than ladder diagrams because programming statements do not look like wiring diagrams.

Ladder diagrams will probably remain the most popular programming method because the users (installer, operator, maintainer, troubleshooter) are not computer programmers. Engineering personnel can possibly use any programming method available, but factory personnel are likely to prefer the ladder diagrams.

NETWORK OPERATIONS

Most programmable controllers have the capability of operating with a supervisory computer, which collects data from many controllers and in addition, may direct operation of the programmable controllers. An overall production line can thus be coordinated. The supervisory computer coordinates several machines and parts flow and collects data on the performance of the individual machines. The supervisory computer may have stored all programs needed for the individual controllers and "download" those programs to the controllers when needed for a product change. The programmable controllers can collect and store any desired data, in addition to the control functions; the supervisory system computer then can request those data, hourly or daily, as required. The supervisory computer can present data summaries for management purposes: machine performance, product yield, percent rejects, downtime, or any other conceivable data.

Additional benefits are as follows:

1. Greater reliablity due to the elimination of most relay contacts. Contacts are subject to arcing and sticking; this is particularly important in dirty locations.
2. Less space required for the programmable controller than for relay panels.
3. Economy: Programmable controllers are available which show an economic advantage even when replacing as few as 4 to 10 relays.
4. Data collection: As part of their normal functions, programmable controllers can collect such data as machine operating times and parts tolerances.

DIAGNOSTICS

A recent improvement in programmable controller capability is machine diagnostics. When a machine breaks down, most troubleshooting time is used in finding the fault, whereas the actual repair may be fairly quick. Programs are available that monitor times required for each machine operation or dimensions of parts produced, and an alarm is sounded if these variables change beyond the tolerances prescribed. This feature can be of considerable help in troubleshooting, when a machine shuts down and several alarm lights come on. The diagnostic program can tell you what alarm was first and give you a good start in the right direction toward finding the fault.

CONSTRUCTION

Specifications are discussed in Chapter 11 and also Chapter 6; however at this point a strong reminder is important: Be sure to specify rugged industrial equipment. The factory environment is dirty and hot, subject to vibration, and contaminated with radio and electromagnetic interference. A programmable controller must be designed and built to withstand these extreme operating conditions; the usual construction techniques for office machines (to be used in a clean, cool atmosphere) are not adequate for the factory environment; therefore, be wary of equipment that is basically office equipment repackaged in a different box, which looks rugged.

The I/O modules are especially critical, in that good design in mandatory to avoid problems with ground loops and electrical noise. The use of optical isolators is highly recommended, as discussed in Chapter 6 and Appendix B. A typical industrial-quality I/O rack is shown in Fig. 10-9. Two types of modules are shown: 115-V ac/dc inputs and 115-V ac outputs. These are typical on/off signals to and from switches, pushbuttons, relays, lamps, or motor controls. The modules plug into the back plane of the rack. Field wiring connects to screws in the hinged blocks at top and bottom (the top one is shown hinged up). Field wiring does not have to be disconnected when modules are changed (see Fig. 10-10); simply swing out the connection blocks and pull out the modules.

An example of good rugged printed circuit board construction is shown in Fig. 10-11; this is a CPU module. The printed circuit (PC) board is securely attached to the steel front panel; the PC plug-in fingers (at the top) are heavily plated and widely spaced to minimize the possibility of shorts due to dirt and moisture. Circuit traces are widely spaced so that visual inspection for faults is easy; soldering is neat, with no "tails." Plug-in cable connections (at bottom) are securely fastened so that in-

Fig. 10-9 Programmable controller rack. (Courtesy of Honeywell Industrial Programmable Control Division.)

Fig. 10-10 Programmable controller I/O rack. (Courtesy of Honeywell Industrial Programmable Control Division.)

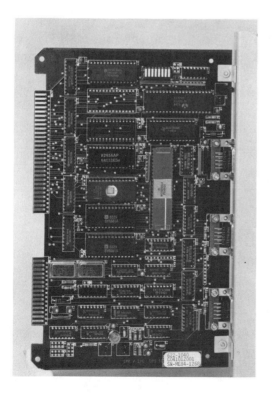

Fig. 10-11 Programmable controller CPU board. (Courtesy of Honeywell Industrial Programmable Control Division.)

serting and removing external cables does not break solder joints on the board. Not obvious in the photograph is the quality of the circuit board stock: fiberglass base for strength and resistance to cracking. Flexible PC boards will crack, causing breaks in wiring traces.

SUMMARY

The uses of programmable controllers are increasing rapidly, from machine control to bakeries, to chemical batching, to papermaking. Originally designed as a relay replacer, the typical programmable controller today includes functions such as data collection, report generation, and network operation.

The advantages are many. The chief precautions are to (1) specify rugged industrial equipment, (2) install it properly, and (3) provide adequate training for operators and maintenance personnel.

STUDY QUESTIONS

1. Outline several advantages of programmable controllers.
2. Describe the basic operational block diagram of a programmable controller.
3. List important considerations when selecting a supplier.
4. What are the advantages of optical isolators?
5. What is a relay force command?
6. How is a wiring diagram different from a ladder diagram?

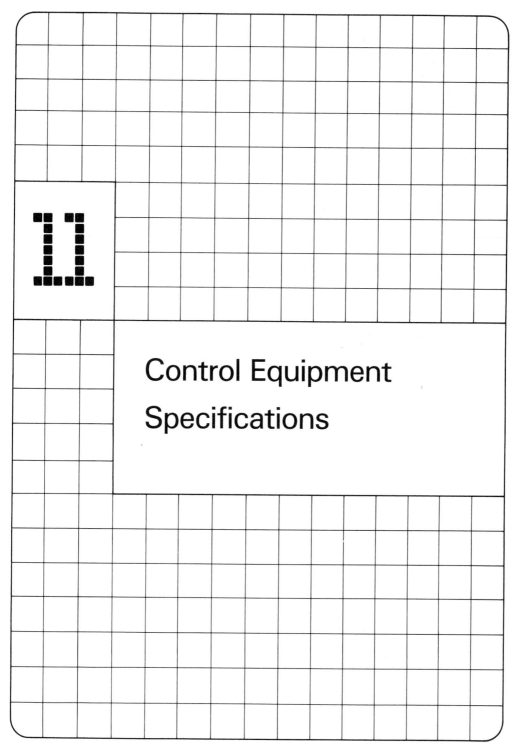

11

Control Equipment
Specifications

OBJECTIVES

To provide guidelines for writing control system specifications

To discuss topics frequently overlooked in specifications

To suggest guidelines for evaluation of vendors of control equipment

In this chapter we discuss items frequently left vaguely defined or omitted entirely. Authors of control equipment specifications should consider the items listed, in addition to the usual material included in a specification.

Although an engineer will usually be responsible for writing specifications, an experienced technician can contribute much when specifications for new equipment are being prepared. After all, the technician has valuable experience in practical matters of maintenance and troubleshooting and knows the weak points of existing equipment. Past mistakes in selecting control equipment may not be repeated if the technician is consulted before purchasing new equipment.

COMPONENT OR SYSTEM SPECIFICATION

The first decision to make is whether to write a system specification or a series of component specifications. In the first case, if a system specification is to be written, the general functional requirements must be defined. These requirements must be defined as exactly as possible: just what the system must do and under what conditions. The selection of sensors, controllers, and actuators is then the responsibility of the vendor, who must supply an operating system meeting the overall requirements. (But even in a system specification, many details must be specified in order to obtain a satisfactory control system which is easy to set up, is reliable, and is easy to maintain.) Many things are left to the vendor's discretion in a system specification, and therein is a possible pitfall. Such things as drift allowable, maintenance provisions, and so on, must still be defined by the user. The vendor then assumes responsibility for detailed design and compatibility of the various parts purchased, such as motors, amplifiers, valves, and measuring instruments. It is the vendor's responsibility to specify operating requirements of each component such that the overall system will meet all requirements. The worst-case tolerances of each component must be evaluated to ensure that the overall system will operate satisfactorily over the required range of temperature, humidity, vibration, supply voltage and frequency variations, and so on.

   If a series of component specifications is to be written, it is the user's responsibility to write a detailed list of requirements for each part. It is quite a task to specify and purchase pieces of a control system from several vendors and assemble a system. Each vendor must be studied separately because each is likely to have different definitions and tolerances on signal levels, impedances, environment, and dimensions. The only way to assure some degree of success is to provide sufficient time in schedules so that each vendor's product can be thoroughly tested before an attempt is made to fabricate and connect the entire system. If a dozen pieces of just-received equipment are connected and the system does not perform correctly, then piece-by-piece testing must be done, probably on a rush basis, to find out what piece is below specifications and/or incompatible. It is far better to provide for thorough testing and data taking as the pieces are received. Debugging of the overall system is much easier if complete data on each component are available for on-site testing.

   The choice of whether to write a system or component specification depends on the user's knowledge of process detail. The user should be familiar with the process and should be the best qualified to specify control system details. It is generally better for the user to learn controls than to try to get a control vendor to understand the process to be controlled. There is also the problem of possibly secret details of a process, which you might not want to explain to an outside vendor. How can an outsider design a successful control system for you if you refuse to explain everything that goes on in the process?

TYPE OF APPLICATION

A specification should explain the application: military, industrial, laboratory, test facility, or whatever. The category will have great effect on the type of construction and type of maintenance provisions. An experimental facility may require a very flexible control system which is easy to change. On the other hand, an industrial control system on a production line must be very reliable, easy to set up, hard to misadjust, and easy to maintain, so as to minimize downtime. A control system for a laboratory facility would not need the extreme reliability that is necessary for an industrial process; downtime would not be so troublesome.

ITEMS TO BE SPECIFIED

Accuracy

Control accuracy should be specified in either a component or system specification. For example, for a specified combination of input signals, the output signal should be within some tolerance, say 2%, of the calculated value, for all conditions of line voltage and frequency variations and temperature.

Dynamic Response

The required speed of control equipment should be specified; this may be valve actuator speed and step response as well as bandwidth and transient response of an electronic amplifier. It is also wise to put in a requirement on time to recover from overloads, because some equipment may "get stuck" in overload conditions following large transients—unless designed to avoid such problems. The dynamic response required can be determined by means of a simulation study in which the performance of simulated components with differing characteristics can be studied.

Overall Steady-State Gains

The simulation study will also provide good estimates of the controller gains required—such as how much output signal is needed for a given temperature error. Experimental data from the process should also give some indication of the gains required. If these gains are not specified, the vendor can only guess and may supply too much or too little gain.

Control Adjustments

A control system specification should define which controls must be fixed and which are to be adjustable. In addition, the range of adjustments should be specified. Again, the simulation study can furnish this information. If not specified, a gain adjustment may be supplied with, say, a range of 0 to 500, compared to an actual intended range of 10 to 20. The resolution and ease of adjustment suffer badly using the wide range control. (How would an operator set the gain to a value of 17, for example, on the wide range control that covered a range of 0 to 500?)

A minimum of knobs should be specified and no extras allowed. Extra knobs provide flexibility, but also make setup difficult and increase the probability that an adjustment will be changed by mistake. Adjustments that are seldom used should be screwdriver adjustments, not knobs. Note that as emphasized elsewhere, all control settings must be recorded, so that if settings are changed (perhaps hit accidentally, or changed deliberately by someone with good intentions), the controls can be set back to the original values.

In this same category, the power on/off switches should not be on the front panels, or at least not prominent, where they can accidentally be moved to "off." Similarly, if power cords are plugged into receptacles, clamps should be installed to hold the plugs in.

Environmental Sensitivity

A specification should include all known conditions that are important; these include temperature range, humidity, cleanliness (or dirtiness), supply voltage and frequency variations, shock and vibration, and others. Requirements should state, for example, that a line voltage change of $\pm20\%$ should cause no more than a 2% error in controller output.

In addition, it may be wise to include a requirement that "spikes" and other rapid transients in supply voltage shall cause more than a 2% error in controller output. There have been cases in which a power supply performed well for slow variations in line voltage, but actually amplified spikes (see Chapter 6 for comments on this phenomenon). These spikes may cause extra inputs to the control system, causing mysterious changes in set points and false actuator movements, which can cause errors in machine operations.

Do not overstate requirements, however, because the cost of equipment increases very rapidly with stringent operating conditions. If equipment is to be installed in an air-conditioned control room, the temperature and humidity variations would normally be very small compared to an outdoor installation. There would be no need to require equipment that is waterproof, or an operating temperature range of 20 to 150°F.

Radio-Frequency Interference and Static Discharges

These may be a serious problem. As an example of specifications for good industrial equipment, consider the following (relative to the Honeywell UDC-5000 Universal Digital Controller, discussed in Chapter 9).*

*These descriptions for interference susceptibility are excerpted from specifications for the UDC-5000 Universal Digital Controller, courtesy of Honeywell Industrial Controls Division.

Static charge susceptibility. The exposed panel surface is capable of withstanding a discharge through 100 Ω from a 250-pF capacitor charged to 10 kV with no component failures and 8 kV with no incorrect outputs.

Radio-frequency interference (RFI) susceptibility. The Universal Digital Controller is capable of withstanding an RFI field strength of 20 V/m at 27 MHz and 151.685 MHz with an analog effect no greater than 2.0% and otherwise normal operation and at 450 MHz with an analog effect no greater than 4.0% and otherwise normal operation.

Noise Output

A specification should state the maximum permissible signal noise generated within the equipment. For example, a 10-V controller output signal might be specified to include not more than 10 mV of noise. This limit would be set by the allowable "jitter" in an actuator, caused by the noise—which is unacceptable due to unnecessary wear and/or heating of the actuator motor.

The noise referred to is that internally generated in the control equipment; it is measured with all input signals disconnected and input terminals grounded. All gains and adjustments are set at typical values (not zero!).

Another type of electrical noise which is very difficult to limit by a specification is that which comes out of equipment on the power line. For example, switching-type power supplies use internal square-wave oscillators which can generate considerable electrical noise both radiated and conducted out on the line input wiring. Vendors should be asked about this, to determine their depth of concern for your problems.

Drift Allowable

It is vital to specify how long the equipment must operate before readjustment is required. This is a matter of stability: drift in gain settings, set points, and balance adjustments. These are internal equipment design problems and are in addition to problems caused by supply voltage variations, for example. It seems reasonable that a control system for an industrial process should operate for at least a week before requiring readjustment due to drift.

Input/Output Connections

The specifications must include complete information as to sources of signal inputs and destinations of signal outputs. This would include voltage, amperage, impedance levels, bandwidth of signals, noise level, type of cable connectors, and type of cables required. These considerations are aimed at minimizing problems of incompatibility of the several vendors' equipment.

Factory Tests

Each specification should include a requirement for a full factory test and demonstration of the equipment before acceptance and shipping. The tests must be done according to a procedure agreed to before testing, aimed at demonstrating compliance with all specification requirements. These tests also supply data useful for troubleshooting later; these tests involve setting up simulated input signals to the system and recording system outputs. A complete run-through of alignment procedures is done. Dynamic as well as static measurements should be made as well as determining sensitivity to line voltage and frequency. In some cases, a regular vendor factory production-line test may be accepted in place of some test. The test program should fully demonstrate that the equipment performs according to specifications. A simple statement by the vendor "that the equipment is guaranteed to work" is not acceptable.

After installation at a site, it is very awkward and expensive to reengineer a faulty system, *no matter at whose expense.* Any troubles will reflect on the design engineer and the technicians, not to mention the expense of lost time while numerous equipment difficulties delay other portions of a project.

Maintenance Provisions

It is highly desirable to build in maintenance provisions such as test points, test meters, and voltage test sources. The extra cost is more than offset by savings in time when troubles do develop. A technician does not have to connect several external meters and external voltage sources in order to diagnose troubles. Another highly useful feature is monitor meters built in so that technicians can observe performance of the equipment without having to connect external meters. These internal meters could show continuously the error voltage and actuator demand signal.

Failure monitors are also useful for indication of excessive signals, or lack of signals. They can be wired for automatic switching of the control system from automatic to manual, thus avoiding disturbing the plant with faulty control signals.

The features listed above are very useful when ease of operations is important and downtime must be minimized. The extra cost incurred must be justified for each control system design and that depends on the nature of the applications.

Guarantees

A meaningful gurantee is hard to get, but if a vendor has system responsibility it seems reasonable to require a guarantee of successful operation at the installed location. If troubles develop, a vendor may charge hundreds of dollars per day while correcting their errors in design. A thorough factory test, will, of course, reduce the number of troubles appearing after installation at the site.

A guarantee should also be required stating that the equipment will operate according to specifications for some minimum period of time (after installation, not after shipment from their factory). Ninety days is probably a typical figure. Some vendors give free service for long periods just to maintain their good reputation. Along with the foregoing, a penalty clause should be included in any specification, requiring the vendor to pay a fee if the promised delivery time is not met.

Hardware Details

Sometimes there is a preference for a certain type of equipment, such as pneumatic, hydraulic, or electronics with transistors, integrated circuits, or magnetic amplifiers. If so, that preference must be stated in the specification and a vendor selected with known capabilities in that type of equipment.

There may be preferences in cabinet size, shape, material, finish, and so on. Also, drip-proof cabinets may be necessary, with lifting hooks on top or forced-air ventilation. Is cable entry to be from top or bottom? All fuses and circuit breakers should be on the front panels if at all possible. All of these particulars should be specified to avoid controversy later (price increases, reengineering costs, and shipping delays).

Instruction Books

If a vendor has system responsibility, the specification should call for instruction books explaining theory of operations, tune-up procedures, test equipment required, typical operations, troubleshooting procedures, and maintenance procedures. A good instruction book is custom-prepared for a particular system and is not just a collection of catalog sheets for various components.

The instruction book must include a suggested list of spare parts and particularly important, a list of commercially available components, such as transistors and integrated circuits, which are equivalent to specially numbered vendor parts. Vendors may balk at this, but it is worth fighting for. Many equipment manufacturers buy standard parts and put their own special part numbers on them. This practice can make maintenance a real headache, because if you are short of spare parts, it is difficult to find out what standard commercial parts might work: for example, transistors and integrated circuits. Many commercial parts may work, and at a lower price. Always ask to see typical instruction books prepared for other customers on jobs of similar complexity.

SELECTION OF CONTROL EQUIPMENT VENDORS

System or Component Supplier

Vendor selection depends greatly on whether a system is being purchased, or whether individual components are being selected and purchased by your company. The component supplier may ship off-the-shelf components, but a system will require some engineering by you, the user. There are many firms that manufacture good components such as valve actuators, controllers, resistance thermometers, and flow meters. However, there are few firms prepared to supply adequate applications engineering, which involves matching of component requirements and control specifications. Adequate design involves much more than thumbing through catalog sheets!

Specialized Experience

Many vendors have specialized to some degree, with most of their experience in pneumatic boiler control, or data acquisition, or telemetry, to name a few specialities. Obviously, you will get the most for your money and the most reliable system with fewer design errors if you deal with a vendor having much experience in the type of equipment that is being purchased.

Never be the first customer for equipment vastly different from previous products. For example, it is just too much to expect that a mechanically oriented company will become a leader in digital computers overnight. Similarly, you should not expect a first-class valve actuator from a vendor whose whole product line is electronic amplifiers. Also, beware of industrial vendors who state that their manufacturing techniques can be upgraded overnight to meet better tolerances and reliability; quality control procedures may be inadequate or absent completely.

Product-Line Experience

Also determine how long a product line has been manufactured; it takes time and experience to eliminate engineering shortcomings and manufacturing difficulties. For example, the basic modules of an electronic control system may have shortcomings (bugs!) that become evident only after the first few systems are assembled, using dozens of the basic modules. Endeavor to locate and do business with vendors having proven products.

Factory Test Procedures

Always find out what test procedures are used to test modules, cabinet assemblies, and overall systems. Is testing and quality control thorough, or done spasmodically only when required by a customer? Request copies of system test procedures used for jobs similar to the one under consideration. There may be factory tests to observe from which to draw conclusions regarding vendor quality control.

Reputation

A vendor's reputation for quality work, technical competence, reliability, and on-time delivery is best evaluated by talking to other customers. Find out what types of equipment difficulties were discovered after installation, how they were corrected, and at whose expense.

Guarantees

Will the proposed vendor guarantee (for a fixed equipment price) that the equipment will meet all specification requirements on the job site? Whether they will or not is an excellent indication of experience on similar jobs.

A typical trick is to quote a fixed price, plus hundreds of dollars a day for a field engineer "to aid installation;" obviously, this may serve as a cover-up for design errors. The final equipment price may escalate 100% or more by the time the equipment is made to operate according to specifications. A fixed-price proposal thus has no meaning if the customer must pay for correction of vendor errors.

Application Engineering

Some vendors can be of great help in writing specifications. This is the function of the application engineer who studies your problems and is qualified to suggest good solutions and locate potential trouble areas. Good applications engineers can help avoid many problems by virtue of experience on similar jobs. Of course, specifications will "lean" toward their equipment, but that is legitimate salesmanship. You should therefore discuss the proposed system with several competent vendors. There is usually some room for negotiation with potential vendors; perhaps some specification item is too difficult and a slight relaxation would reduce costs considerably. A good field engineer can suggest such changes. A good application engineer knows the capabilities and limitations of the equipment, where it has been installed successfully, and what problems are likely to arise, and be able to discuss intelligently possible solutions in potential trouble areas. The caliber of these application engineers is likely to reflect the quality of engineering at the factory. A competent engineering office is not likely to send out an incompetent sales engineer.

Price

The initial cost of equipment is only one factor in the selection of vendors. Maintenance and downtime costs over a few years may be several times the original equipment cost, especially if the equipment is troublesome and unreliable due to inadequate design. It is usually better to purchase equipment that is somewhat more expensive than the cheapest in order to obtain reliability, ease of maintenance, and long life.

Training Programs

Does the prospective supplier offer training classes suitable for users? Can you attend these classes before purchasing equipment? Are there classes in maintenance procedures?

SUMMARY

The user (you) are the most knowledgeable about process performance and control functions, and therefore should specify many details in equipment specifications. Many of the items discussed in this chapter have been discovered by this writer, the hard way, through painful experience. Don't ever omit a topic from a specification, assuming that any vendor will do things the way you would. Be sure to select competent, experienced suppliers who are interested in helping you achieve good control performance.

STUDY QUESTIONS

1. How can a *systems specification* for a complete control system (sensors, controllers and actuators) lead to difficulties later?
2. Describe how you would evaluate a vendor for product line experience.

3. List several items you should look for in factory test procedures used by possible suppliers.

4. What pitfalls can occur even when a vendor guarantees that your specifications will be met?

5. In evaluating a possible vendor, list several questions that you might ask other customers.

6. Explain how a *fixed price contract* can lead to difficulties.

7. In making a factory tour of a possible supplier, how can you determine the extent of quality control procedures?

8. List several important topics that you should require to be thoroughly covered in equipment instruction books.

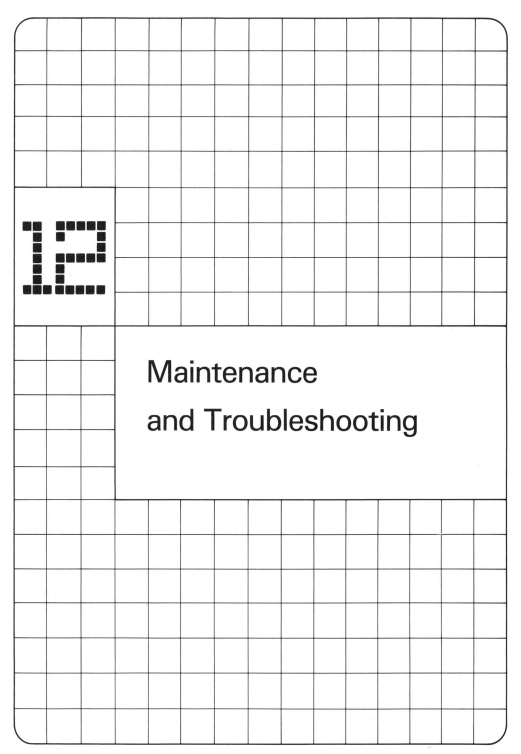

Maintenance

and Troubleshooting

OBJECTIVES

To emphasize the need for adequate test equipment
To illustrate good preventive maintenance procedures
To outline a troubleshooting procedure
To indicate types of faults occurring frequently
To discuss several types of control failures
To emphasize the importance of good record keeping

A good maintenance program begins with good installation, which includes good cabling (rugged and protected), accurate records of wiring, and proper fans and filters. Adequate test equipment must be provided so that faults can be found and repaired to minimize downtime.

TEST EQUIPMENT

Adequate maintenance requires good test equipment; complex electronic equipment systems cannot be properly tested, or faults found, with only the usual volt-ohmmeter (VOM) as provided for industrial electricians. Typical equipment required for a control system using analog and digital signals would include at least the following items:

Electronic volt-ohmmeter. A meter having a high input resistance (over 10 MΩ) as compared to an ordinary volt-ohmmeter (VOM) having an input resistance of 1000 Ω/V is needed. The low-resistance meter may be adequate for checking power supplies, but many electronic circuits would be "loaded down" by such a meter, causing false readings and even causing the circuit to operate incorrectly. Suppose that you want to measure the voltage between points P and Q in the circuit shown in Fig. 12-1.

Figure 12-2 indicates the problem; the true voltage is 31.25 V, whereas with the meter connected, the voltage drops to 22.72 V (and that is what the meter reads). The meter actually changes the circuit, because it is a resistance connected in parallel.

A higher-resistance meter, say 20,000 Ω/V, would read 30.67 V, and a 10-MΩ electronic voltmeter would read 31.19 V.

High-resistance circuits, like the circuit in Fig. 12-1, should be checked with an electronic voltmeter, which would disturb the circuit very little. When making any measurement you must try to estimate how much the meter disturbs the circuit and how much error is caused.

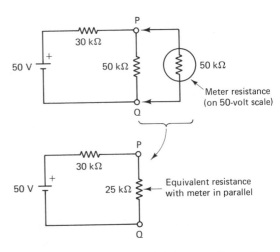

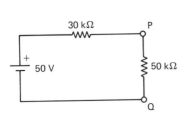

Fig. 12-1 Sample circuit for voltage calculations. The true voltage may be determined thus:

$$V_{PQ} = \left(\frac{50K}{50K + 30K}\right) 50$$

Fig. 12-2 Circuit loaded with voltmeter. The voltmeter reading may be determined thus:

$$V_{PQ} = \left(\frac{25\ K}{25\ K + 30\ K}\right) 50 = 22.73V.$$

Analog or digital meters. Modern digital meters are rugged and give a display that is accurate and easily read—usually. If the measurement is changing rapidly, the blur of numbers is hard to evaluate; similarly, if electrical noise is on the signal, the numbers may shift rapidly. In that case, an analog meter is superior because the moving needle will average out the fast changes and show the true trend of the data; the needle is easily read to determine which way the variable is drifting; thus both types of meters are useful.

Oscilloscope. An oscilloscope having a frequency response of at least 20 MHz, and preferably 60 MHz, with a sensitivity of at least 2 mV per scale division, is needed. A dual-trace oscilloscope is useful not only for observing two waveforms at the same time or measuring phase between two signals, but also for the feature which allows reading the difference of two waveforms; that feature is particularly handy when measurements are required in a circuit such as that shown in Fig. 12-3; suppose it is necessary to observe the voltage across the capacitor. That cannot be done with a single-trace oscilloscope because you cannot connect the oscilloscope ground to point X and the probe to Y (if you do connect the oscilloscope ground to point X, you essentially are connecting a jumper to G, causing a short across the resistor). Some dual-trace oscil-

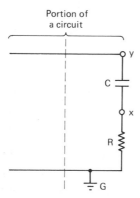

Portion of
a circuit

Fig. 12-3 Circuit measurements with ground difficulties.

loscopes have a difference feature, so that you place one probe on X and the other on Y; the scope displays the difference, which is the desired waveform across the capacitor. Otherwise, the only way to obtain the waveform across the capacitor is to observe the waveform at X, sketch it on paper, observe the waveform at Y, sketch it, then subtract the two sketches to obtain your estimate of the capacitor waveform.

Signal generators. Sine-wave, square-wave, and pulse generators should be available, as specified by the manufacturer of each piece of control equipment.

Voltage and current sources. Sources for generating test signals to controllers and actuators would include 0 to 10 V dc and 4 to 20 mA, and other particular ranges used in your equipment. These signals must be well regulated, free of ripple, not grounded, and adjustable to precise values. Several of these should be available, with some portable units. Possibly these could be built as required, for your specific needs.

Regarding quantities of test equipment: It is always wise to purchase at least three identical pieces of each type of instrument, because test equipment can fail at the most inconvenient times. In addition, if one instrument fails, you may be able to find the fault by comparing voltages and waveforms in a "good" instrument with the faulty instrument. The repair of test equipment can be difficult because of special value components, unmarked parts, and lack of good service information, such as test point voltages and waveforms.

Test Equipment Hazards

Some test equipment and tools can cause trouble, and even destroy electronic components. For example, an ordinary volt-ohmmeter used on the ohms scale has an internal battery which may range from 1.5 up to 30

V, depending on the model. If you are checking a printed circuit board with an ohmmeter, the ohmmeter can burn out transistors and integrated circuits. There are special low-power ohmmeters, using less than 0.6 V, which should be used for measurements in equipment involving transistors and integrated circuits.

Even the ordinary soldering gun (the type with the single-loop high-current tip) can cause trouble. The extremely high current in the tip can induce current in circuitry being repaired, damaging components and creating other faults.

Static electricity is always a hazard because solid-state components can be destroyed by a static electricity discharge which may be several thousand volts. Before using a soldering iron of any type, you should be sure that the tip is grounded, as well as the circuit being repaired.

Power Source for Test Equipment

You should keep in mind that test equipment may be carried anywhere, so where do you plug it in? It should be plugged into the same line that is supplying the electronic equipment on which you are working. The primary reason is to avoid line noise, which may be excessive on lines that supply machinery, and may even cause the test equipment to read incorrectly. In addition, if the oscilloscope or other test equipment is on a different power line from the control equipment, you may introduce noise into the control system, which means that your measurements may be faulty, because the noise may go away when the test equipment is disconnected.

A situation that can be troublesome is indicated in Fig. 12-4. Industrial power distribution frequently uses a four-wire 208-V three phase

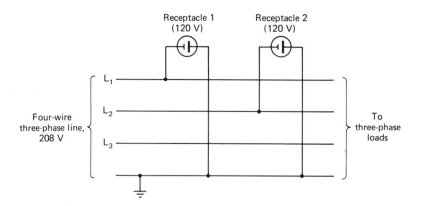

Fig. 12-4   Three-phase power distribution.

system, which means 208 V between line wires and 120 V line to ground. Each receptacle has a "hot" side which is 120 V to ground, but there is 208 V between the hot lines on the two outlets shown! That is a voltage not usually expected between 120-V outlets (you would expect zero volts between the hot sides). It is preferable to have only a single-phase 120-V line to your test bench, particularly, to avoid dangerous surprises.

Another problem, which can be serious, is power-line frequency. Some test equipment may be sensitive to line frequency, so that it does not operate properly on a local emergency generator with frequency swings of 47 to 65 Hz. The best solution is to provide some battery-operated test equipment, including oscilloscopes, so that measurements away from the control room can be made without line power connections. As a precaution you should purchase test equipment that will operate properly with the same line specifications (voltage and frequency) as the control equipment.

PREVENTIVE MAINTENANCE

A good preventive maintenance effort will result in less troubleshooting, because many control equipment failures can be predicted or avoided. A prime rule is to keep electronic equipment clean, cool, and closed. Fans and filters, if provided, must be kept operable, because dirt and heat are major causes of failures. Make a record of equipment control settings and adjustments while the equipment is operating properly; then you can reset the adjustments to proper settings if anyone tampers with them.

Keep an eye out for cable problems; avoid flexing cables wherever possible; for example, if wagons or carts hit a cable frequently, it will not be long before wires inside the cable are damaged, creating a troubleshooting problem. Reroute the cable and fasten it down with clamps before trouble occurs. Also watch for moving parts of machinery rubbing on cables; if a cable connects to a moving part, at least arrange the cable so that it does not drag on anything while moving. Watch for brittle insulation which cracks and falls off when touched; excessive heat or leaking oil or chemicals can cause failure of insulation, and eventually, electrical problems. Replace the cable before it fails (during the next scheduled machine shutdown) or make a guard to keep leaking substances from contacting the cable.

Check electronic equipment frequently and watch for hot spots, such as transformers and resistors causing discoloration of printed circuit boards. Such discoloration may be an indication of a future failure, and could be repaired before further damage occurs.

Avoid vibration if at all possible; control cabinets attached to punch presses, for example, will probably have parts shaken loose and solder joints broken.

Always ask the machine operators to let you know of any peculiar noises or smells. They are very familiar with machine operation and will recognize something abnormal. Squeaking fan bearings may indicate a possible fan failure, with consequent overheating of electronic parts. The fan should be lubricated or repaired before it stops completely. Machine operators might also tell you of excessive sparking of relay contacts, indicating a need to clean dirt from the relay surroundings, and possibly to polish up the contacts—before the relay fails or starts a fire.

Lubricate moving parts such as bearings, gears, or sliding surfaces with the proper lubricant; use proper-weight oils in bearings, and grease on sliding surfaces; oil runs off, particularly at high temperatures; do not use so much that excess drips on electrical wiring or components.

Replace indicator lamps when they fail; if lamps fail too often, you can make them last longer simply by putting a resistor in series, to reduce operating voltage to 100 V on a 120-V lamp. Lifetime will be greatly extended by that reduction in voltage, and brightness will be adequate.

Power lines. Always check to be sure that a separate power line has been installed for electronic sensing equipment and control cabinets. A main power line to a machine will have excessive variation in voltage and noise spikes, which may cause mysterious troubles with controls. That sort of trouble is very difficult to locate, because it does not repeat often enough for you to locate the source. It is cheap insurance to provide a separate ac power line in a separate conduit to the control equipment. To avoid further difficulties, it is wise to install surge suppressors on each ac power line, in each control equipment cabinet; these suppressors will greatly reduce any voltage spikes that do arrive.

TROUBLESHOOTING

There really are two very different types of troubleshooting:

1. When the plant and controls are first installed and started up
2. Later in plant operation when something fails

If the first case, you really do not know if the plant wiring is correct; you do know, hopefully, that each control component (sensors, controllers, and actuators) does operate, because each should have been checked out (by you) when it arrived at the plant site. At least during initial startup of a plant, everything is relatively clean, unless cabinets are left open unnecessarily.

Later in plant operation, you can generally assume that plant wiring is correct, because everything has been running. Even then, some original equipment defects may not show up for 10 to 12 months: for example, cold solder joints, which operate satisfactorily until atmospheric conditions cause slight corrosion between parts of solder joints, causing an intermittent contact.

Wiring Drawings

Regarding plant wiring: Sometimes equipment is not connected according to installation drawings, but somehow everything works. For example, someone ignores a color code or wire label specified for control wires CS-11 and CS-12 in a cable. Everything is hooked up, but with CS-11 and CS-12 reversed on both ends of the cable, which of course makes the system operate correctly. The difficulty appears when the system breaks down for the first time; a technician finds CS-11 and CS-12 reversed in a cabinet—then what? If the wires are switched now (on one end), a bigger problem is caused because things were connected correctly in the first place, but not according to drawings.

The point is, wiring diagrams should show exactly how everything is connected and show any changes from original installation drawings. In the example above, if someone had simply marked the drawings showing reversed wires CS-11 and CS-12 to show how installed, there is a record of what connections worked correctly.

Naturally, it would be preferable to change the wires to agree with the original drawings, but if that simple change would require a plant shutdown, obviously it is better simply to mark the drawings to show as-built connections. Then when it comes to troubleshooting, the technician will have correct as-built connection diagrams—otherwise, much time is wasted ringing out cables to get information that should have been available from the drawings.

Equipment Modifications

On this same subject, if plant wiring is modified, an accurate record should be made to avoid difficulties later. For example, suppose that a dc power supply fails and a substitute is installed quickly to get the plant back on-line. The substitute supply may have adequate voltage and current ratings, but different terminal block connections. The cabinet drawings should be marked to show that the substitute power supply requires wires CS-11 and CS-12 to connect to TB1-18 and TB1-19 instead of to TB1-7 and TB1-8 as shown on the original drawings (the dates and the name of the person making these repairs should also be recorded). Good records of plant wiring are a valuable time-saver when troubleshooting is required.

A good technician will keep a record book showing any wiring changes made, any equipment substitutions, and any measurements made. During initial startup of a plant, records should be made of "normal" voltages and waveforms on every available wire, so that troubleshooting is easier when something does not operate properly.

Any wiring changes should be sent to the plant engineering department or whatever department is responsible for record keeping and the updating of drawings. However, it may be some time before drawings are updated and in the meantime, correct information is needed; hence the technician must maintain up-to-date records of wiring changes for immediate use.

Control Loop Malfunctions

Closed-loop control system faults can cause several types of incorrect behavior. A system may be tuned up correctly and operate as desired for months, then begin to show one or more of the following faults:

1. Does nothing: operation fails completely
2. Causes a runaway of a plant variable
3. Operates in a sloppy manner, allowing a process variable to wander beyond specified limits
4. Causes overshoot in a process variable, when the set point is changed
5. Causes unstable behavior, with the process variable cycling up and down continuously
6. Shows erratic behavior: actuator moves strangely, and loop sometimes seems to be "out of control"
7. Operates intermittently: works properly most of the time, but sometimes shows symptoms 1 through 6

Of course, before searching for equipment faults, you always consider that human error might be involved: Someone changed a controller gain (with good intentions, but did not tell anyone else), or someone hit a control knob with an elbow. In any case, you should have all of the correct controller settings in your notebook so that you can check settings and reset if necessary; then if the difficulties are still present, proceed with troubleshooting the equipment.

Fault type 1 is the easiest to find, because it is a definite failure. An internal power supply failed, or a signal is missing due to cable defects, or a transistor or power-line source failed (or is unplugged).

Fault type 2 is also a definite fault, because an actuator demand signal is probably incorrect; signal tracing will allow you to determine where the incorrect signal originated.

Fault type 3 is more difficult to locate; it acts like a closed loop with very low gain; there seems to be some closed-loop control, but it is inadequate. This fault could be caused by internal circuit problems, such as a shorted capacitor on an amplifier output, or a partial short in a cable causing loading of an amplifier output with consequent low gain and output.

Fault type 4 is somewhat difficult to locate because it has the same symptoms as too high a gain setting in a controller; circuit failures such as open resistors (due to bad solder joints) could cause this behavior.

Fault type 5 is similar to type 4, except that the process variable does not "settle down" but oscillates continuously; this fault also appears to be due to excessive controller gain or excessive sensor output.

Fault type 6 is probably caused by poor electrical connections: cables/plugs, printed circuit board sockets, integrated circuit chips loose in sockets, or bad solder joints. Radio signals, such as from nearby radar equipment or even portable radio transceivers (walkie-talkies) too close to control circuitry, can cause these symptoms. Strong radio signals can cause severe overloading in bias circuits, causing a cutoff of circuit output.

Fault type 7 is the most difficult, because the incorrect behavior does not last long enough to locate. Sometimes a closed loop will perform well for months, then do strange things for several minutes: allow excessive error signal with no correction, drive the actuator when no error exists, or do nothing for several minutes, allowing the process variable to drift out of tolerance. About the only way to determine the origin of this type of fault is to connect extra meters to the control circuits, so that when improper behavior begins, you (or a plant operator) could see, hopefully, which signal became improper first. Meters could be connected to the sensor output, the controller output, and actuator input. For example, if you observe that actuator input becomes different from controller output, a bad connection is probably the culprit.

General Troubleshooting

Always ask yourself the question: Why did the failure occur? Sometimes it is due to age; other failures are due to environmental stress.

1. *Age.* Some components fail due to age, particularly capacitors of the electrolytic type. These will dry out over the years and eventually show less capacitance than originally and also excessive leakage current; solid-state components such as diodes, transistors, and integrated circuits may fail because of manufacturing defects such as inside contamination, which eventually causes a failure.

2. *Environmental stress.* The "environment" includes everything from temperature, humidity, vibration, and a corrosive atmosphere, to electrical surges and spikes on power sources and signal lines.

Always try to figure out why a component failed; otherwise, the same failure may occur again—maybe not immediately, but eventually. For example, resistors do not just burn up by themselves. There must be a reason for the excessive current.

One fault can cause several symptoms; a fault in one circuit may cause other circuits to appear faulty. Thus it is wise to make several measurements and then think about what might be the original cause. Usually, the measurement that is the most in error should be investigated first, rather than the ones that are only slightly off.

Think: Will your troubleshooting cause a control system to malfunction? If you are troubleshooting during operation of a machine or process, be very careful not to cause a malfunction, as moving components and wiring can cause a short or broken component. Attaching meters can cause loading of signal lines so that voltage levels are changed. Be careful not to short pins together on integrated circuit and transistors.

A large number of equipment failures are caused by mechanical-type problems: loose plugs, loose printed circuit cards, broken components, broken wires in cables (due to vibration or being hit by something). Before getting out a lot of test equipment, an experienced technician will make a very careful visual inspection of the faulty equipment, using a good light. All cable connectors will be tested for tightness, all screw-type wire connections tighened, all PC boards wiggled to test firmness of sockets, and a search made for foreign matter such as dust, water, metal shavings, nuts and bolts, and tools left in cabinets (that is one reason that electrical equipment cabinets should be kept closed). Using a soft, clean brush, dirt and trash can be carefully removed (if power is on, the technician must know what voltages are present and where). If only low voltages, under 40 V, are involved, cleaning may sometimes be done carefully with the power on. Otherwise, it is best to deenergize equipment if extensive cleaning is required.

Signal tracing.   Sometimes when tracing a signal from one circuit to the next, confusing symptoms may occur. Figure 12-5 indicates a possible system in which a thermocouple millivolt signal (signal 1) is input to circuit A, amplified, and transmitted as signal 2 to circuit B for comparison to a set-point voltage, signal 3.

Suppose that signal 2 is incorrect. You cannot definitely conclude that circuit A is faulty, because the signal may be incorrect because of a fault in circuit B. The effects of a fault can reflect backwards, not necessarily in the direction of signal flow, which is left to right. The cable carrying signal 2 should be disconnected in order to measure the signal out of circuit A; then you would know if circuit A is defective or not. However, when disconnecting cables, dummy loads are usually required to maintain the proper load on the driving circuit (see Fig. 12-6).

Dummy loads.   Dummy loads are generally required; otherwise, your measurements on disconnected cables may indicate different values of circuit voltage or current than when connected. On ac and dc voltage

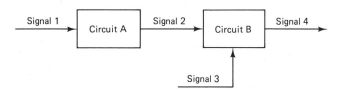

Fig. 12-5   System signal tracing.

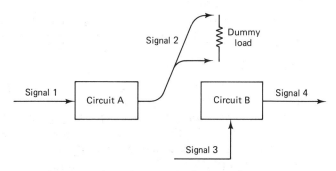

Fig. 12-6   Use of dummy load in signal tracing.

signal lines without a dummy load, measured voltage will read too high. On pulse circuits, reflections and ringing may occur due to lack of proper termination, and your oscilloscope pictures will indicate improper pulse shapes. This manufacturer's maintenance manual should indicate what load should be connected to each circuit output so that you can test a unit when not connected to the next unit. Otherwise, how can you tell if a measurement indicates a fault or if the test equipment is causing the faulty measurement?

If faulty printed circuit card connectors are suspected, turn off power before removing boards. (Manufacturers generally advise against un-plugging and replacing boards with power on because electrical surges may damage components.) Plug-in contacts on printed circuit boards should be bright and smooth, not discolored, oxidized, or pitted. If nec-essary, clean contact fingers according to manufacturer recommenda-tions; if there are no recommendations, use a pencil eraser to polish the contact fingers. While the boards are out, inspect each for broken or missing parts, discolored parts, capacitors leaking, bad or broken solder joints, or other obvious faults.

Circuit boards in correct slots.   Sometimes it is wise to check that printed circuit boards are plugged into the correct places, especially if the equipment has been out of service for a while; someone else may have removed printed circuit boards and replaced them incorrectly.

Power sources.   The first thing to do when equipment operates incorrectly is to check the sources of operating power. (Sometimes you will find that the on/off switch on the front panel is turned off. That is a real hazard, and perhaps those switches should be taped so that they cannot be moved easily.) Start with the ac power line and verify that proper line voltage arrives at the control equipment cabinets; check fuses in the cabinets. Next check the output of the various power supplies in the equipment cabinets, such as +5 V, +12 V dc, and 24 V ac at the power supply output terminal strips. Next check the power terminals at the different chassis in the cabinet and at the printed circuit boards. These meausrements will indicate blown fuses, broken cable wires in the cabinet, loose plugs, loose printed circuit cards, and similar faults.

If all power sources appear proper, use an oscilloscope to check wave-forms on the several dc power lines; that may seem to be a strange mea-surement, because the oscilloscope should show no waveforms and only a straight line dc; however, many types of power supply failures will cause pulses or spikes to appear on the dc lines. Capacitor failure is one such condition; the dc line will no longer be pure dc (although a voltmeter will indicate no problem); besides allowing excessive ripple on the dc line, such a failure may allow pulses from one control chassis to be passed to another control input circuit, improperly, by way of the dc power line. Such "cross coupling" may be very difficult to locate and prove. An oscilloscope measurement on the dc lines is a good precaution if the control system behaves strangely or erratically.

After all power sources have been checked, signals in and out of the control equipment should be checked. Find the signals from the various sensors—thermocouples, tachometers, and so on—and verify that voltage levels are proper. Check controller output signals to the actuators, and compare to "correct" values (in your notebook from earlier measurements, and from vendor instruction books). If the controller appears to be at fault, arrange some voltage sources to represent the correct inputs (simulated inputs) and trace signals through the control equipment to find the point when improper signals begin; at that time maybe a swap of printed circuit boards may be tried, provided that spares are available. An oscilloscope is useful in this signal tracing, to locate noise and spikes on the control signals which may be due to improper cable and equipment shielding and grounding.

Board-Level Troubleshooting

Board-level troubleshooting involves troubleshooting and repair of printed circuit boards after you decide that a fault is on a board. It is generally not wise to attempt board-level repair unless adequate maintenance information is available, including schematics with test point voltages and waveforms, and unless proper test equipment is available as well as replacement parts. If you do not have adequate schematics, it may be very difficult to interpret measurements and decide what is correct and what is faulty. Obvious failures, such as burned resistors or melted capacitors, might be replaced without really knowing their function, but then you do not know what caused the part to fail. A replacement part will probably fail, and you still do not know what caused the overload. Similarly with fuses: Sometimes these will fail due to old age or excessive heat, but usually there is a genuine electrical overload that caused the fuse to blow.

When trying to repair a PC board, how do you test components? An ohmmeter can be used to check resistors for proper values and capacitors for shorts. However, it is generally necessary to disconnect one end of the component to avoid "sneak circuits" which are in parallel with the component. As mentioned elsewhere, it is wise to use a low-voltage ohmmeter to avoid damage to solid-state components.

Making repairs.   Once the defective part is located, it can be replaced, provided that you have the proper replacement part and the proper tools, use proper procedures, and can do a high-quality repair. This means good soldering and no damage to the PC board and other components!

After repairs are completed.   Your job is not finished when you replace the defective part and observe that the system operates properly. The system must be checked for proper calibration and alignment according to the manufacturer's instructions. Data should be taken to verify that all variables involved in the repaired circuit are now correct.

Replace covers, doors, guards, and so on. Fasten down cables and replace all screws and anything else removed during troubleshooting, including PC board hold-down clamps. Do not leave a rat's nest of loose wiring; replace everything in its original condition insofar as possible.

Before leaving the equipment, check the operation of fans if installed, and replace air filters if used. Clean out dust and dirt around the equipment. Check for loose parts, nuts and bolts, and tools left in cabinets, which could fall into equipment and cause trouble later. Make a record of parts replaced and any modifications done, wiring changes, and final measurements verifying that the trouble is corrected; mark maintenance books and schematics showing parts replaced with the date and your name. Replace maintenance manuals and schematics in their proper storage locations. Don't just leave them on top of a cabinet somewhere.

SUMMARY

A good maintenance program will help you to keep the plant running and minimize downtime when something does fail. Proper test equipment and good training are certainly vital. Troubleshooting requires a good knowledge of both the process and the control equipment, because symptoms in closed-loop systems can be very confusing. To avoid wasted time, the technician must analyze what is happening to cause improper behavior. Be careful to make repairs using proper parts, tools, and procedures and good workmanship, including soldering.

STUDY QUESTIONS

1. What are the first items to check when a control system operates incorrectly?
2. Why would you check a dc line with an oscilloscope?

3. What advantage does an analog meter have over a digital display meter?

4. Describe some test equipment problems that may be caused by line-frequency variations.

5. Why connect test equipment to the same power line as that used by the equipment under test?

Appendices

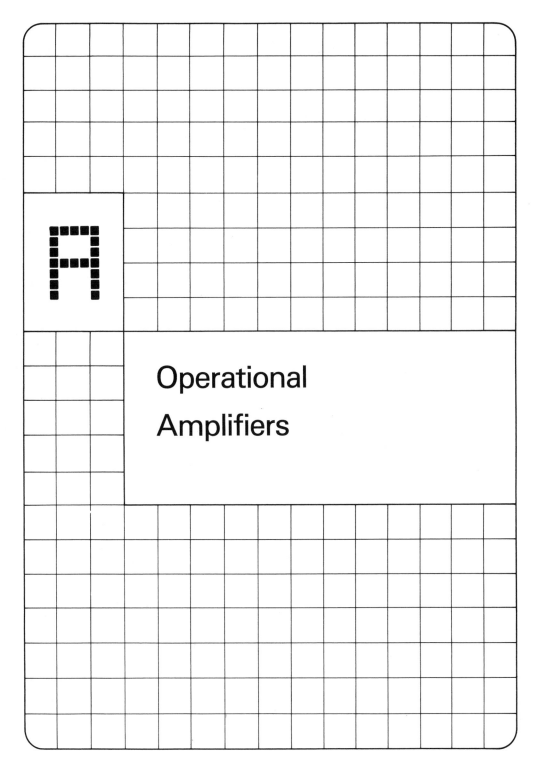

A

Operational Amplifiers

223

Operational amplifiers, called *op-amps*, are an entire subject in themselves and only the basic ideas will be presented here; for a detailed and precise analysis, the reader is referred to the references in the Bibliography.

The op-amp itself is an integrated circuit packaged as a chip, similar to integrated-circuit chips, or else in a round can as are transistors; the circuit is quite complex, usually involving 20 to 40 transistors. A typical inexpensive type is the 741, which has a nominal open-loop gain of 200,000, with a minimum of 50,000.

A basic use of op-amps is indicated in the circuit in Fig. A-1, which shows an inverting amplifier. The purpose of this circuit is to obtain a gain, defined as the ratio of voltage out to voltage in, exactly as gain is defined for any amplifier. This particular circuit could be a part of a controller having proportional control action; then one of the resistors would be variable, to allow adjustment of the proportional gain.

The overall gain of this circuit depends only on the values of the resistors (R_F and R_{in}); the importance of this statement is not obvious until you compare the op-amp circuit to an ordinary amplifier with the same gain. For example, suppose that an amplifier is required having a gain of 10, as indicated in Fig. A-2. The resistor values must be selected for proper bias and a linear operating range of the transistor so that gain will be constant over the desired input voltage range. Now if an application required a gain of 10, the gain of the transistor itself must be known in order to calculate values of the resistors. This can be done, but with difficulty, particularly because transistor gain tolerances may be ±20%. Thus you either test each transistor for gain before designing the circuit, or else you put in adjustable resistors for everything, so that gain .

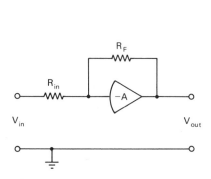

Fig. A-1 Basic operational amplifier. Gain $= V_{out}/V_{in} = -R_F/R_{in}$.

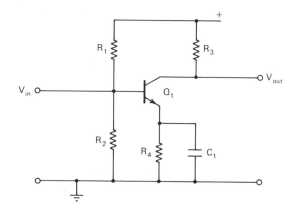

Fig. A-2 Conventional transistor amplifier circuit.

225

can be set to the desired value of 10. (Of course, it is possible to design an amplifier circuit with a lot of negative feedback, internal, so that parts values are not too critical—but this is complicated and really approaching the whole idea of op-amps, using a high-gain amplifier with a lot of negative feedback.)

But assuming that the simple circuit is built with several adjustable resistors (although expensive, time consuming, and generally considered poor design), what happens when the transistor is damaged and must be replaced? A transistor off the shelf may have a gain 40% different from the original one (if the original one was 20% high in gain and the replacement one is 20% low). The circuit must then be adjusted to a gain of 10 again. These adjustments can be complicated if several adjustable resistances are provided because the transistor bias must be set properly at the same time that a linear operating range and proper gain are obtained.

These adjustments are generally impractical for a factory maintenance department. If more than one transistor is used in the amplifier (which is likely), the adjustment procedures become quite complex.

When op-amp circuits are properly designed, the circuit gain depends only on the ratio of the resistors, R_F and R_{in}. The amplifier gain itself, A, can be any number, as long as it is large, say greater than 50,000. Thus if the op-amp chip is replaced, circuit performance will be the same as before, because the resistors are the same. If 1% or better resistors are used, the circuit gain is known very accurately, without knowing exactly the gain of the op-amp itself. The manufacturer's specification will state that the amplifier will have a guaranteed minimum open-loop gain. Above that value, the circuit will operate properly.

In Fig. A-1, A is the open-loop gain of the amplifier, without feedback resistors. The negative sign simply indicates that the amplifier inverts the signal (180° phase shift).

In summary, the op-amp circuit shown provides a closed-loop gain which is simply calculated by the ratio of resistors, and the open-loop gain of the op-amp chip itself does not matter. That is really a marvelous result. Even the design is easier than that for a conventional amplifier, and maintenance is easier because any replacement amplifier (with the same type number) will work as originally intended.

How does this happen? A bit of analysis will be shown here, to indicate how. An "ideal" op-amp would have the following characteristics:

1. Infinite gain
2. Infinite input impedance
3. Zero output impedance
4. Infinite bandwidth

Of course, nature does not allow us to reach either infinity or zero, so practical parts can only approximate the ideals listed above. For practical purposes, a gain of 10^4 to 10^6 is essentially infinity. An infinite input impedance indicates that the circuit would require *no* input current. Actual devices will require microamperes, or nanoamperes, which is essentially zero current for practical purposes.

Zero output impedance means that the circuit can maintain a constant output voltage no matter what load is attached. Practical units can approximate this requirement up to some output current limit. Infinite bandwidth means that the unit will amplify all input frequencies from dc to infinity. Practical units will handle dc to many megahertz, which is quite adequate for most industrial applications.

ANALYSIS

A basic analysis of the op-amp circuit will be shown. Of course, for design purposes, this is too simplified and other matters, such as null adjustments, stability, and finite input current (not zero), must be considered.

Consider the circuit shown in Fig. A-3; the following variables will be used:

V_{in} and V_{out} are the circuit input and output voltages, respectively.

i_{in} is the input current to the overall curcuit.

V_S is the input voltage at the op-amp terminals.

i_S is the input current to the op-amp internal circuitry.

i_{out} is the assumed current through the feedback resistor.

Then the current through resistor, R_{in}, is

$$i_{in} = \frac{V_{in} - V_S}{R_{in}} \tag{1}$$

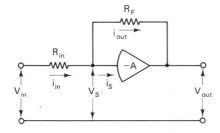

Fig. A-3 Circuit for simplified analysis of operational amplifier.

and the current through resistor R_F is

$$i_{out} = \frac{V_S - V_{out}}{R_F} \qquad (2)$$

Now the current, i_S, into the op-amp input may be assumed to be zero, due to the high input impedance; thus

$$i_{in} = i_{out} \qquad (3)$$

Substituting equations (1) and (2) into equation (3) gives us

$$\frac{V_{in} - V_S}{R_{in}} = \frac{V_S - V_{out}}{R_F} \qquad (4)$$

Then

$$R_F V_{in} - V_S(R_F + R_{in}) = (-)R_{in} V_{out} \qquad (5)$$

Next we substitute for V_S. Since

$$V_S = \frac{V_{out}}{(-)A} \qquad (6)$$

equation (5) becomes

$$R_F V_{in} + \frac{V_{out}}{A}(R_F + R_{in}) = (-)R_{in} V_{out} \qquad (7)$$

Then solving for the gain yields

$$\text{gain} = \frac{V_{out}}{V_{in}} = (-)\frac{R_F}{R_{in}}\left[\frac{1}{1 + \underbrace{\frac{1}{A}\left(\frac{R_F}{R_{in}} + 1\right)}_{\text{second term}}}\right] \qquad (8)$$

Thus if the op-amp gain A is very large, the second term of the denominator is very small, and equation (8) is essentially

$$\text{gain} = (-)\frac{R_F}{R_{in}} \qquad (9)$$

Thus, if A is larger than, say, 10^4 and the closed-loop circuit gain is a small fraction of A, the second denominator term is trivial for all practical purposes. For proper design, the closed-loop gain should be a small percentage of the open-loop gain; otherwise, the second term is not trivial.

To summarize, the overall circuit gain depends only on values of resistors, provided that A is very large and overall circuit gain is not required to be over 100 typically. If more gain is required, another op-amp stage may be added after the first; the overall gain is then the product of the two stage gains. Of course, there is a practical limit to the overall gain that can be obtained by cascading two circuits. That limit is set by the minimum electrical noise at the input of the first amplifier. If that noise signal is 100 μV, the maximum usable gain is probably about 10,000, because 1 V of output signal will be caused by the input noise. For an amplifier operating on a 15-V supply, 10 to 13 V is the maximum output obtainable; thus an input signal of about 1 mV would cause full output if the combined circuit gain is 10,000.

RESET AND RATE ACTION

Other useful op-amp circuits were mentioned in Chapter 3, for reset and rate action. Reset action is obtained by using a capacitor in place of the feedback resistor, and rate action is obtained by using a capacitor in place of the input resistor. Many other functions are possible, such as square root, as required when using differential pressure signals from flow transducers. Detailed information on these circuits may be found in references listed in the Bibliography.

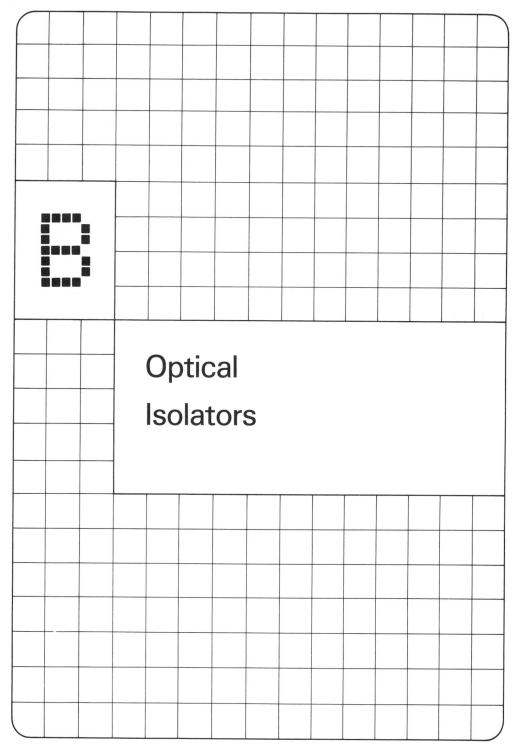

Optical
Isolators

Optical isolators (optocouplers) are probably one of the most useful devices since the invention of the integrated-circuit operational amplifier. Optocouplers consist essentially of a light source (usually an LED—light-emitting diode) and a photosensitive device such as a phototransistor or photodiode. A schematic of a typical device is shown in Fig. B-1.

Ground loop problems between racks, especially between control room and remote electronic equipment, can be eliminated because a separate power supply can be used to power the two sides of the optocoupler: one on the signal input side and one on the signal output side, with no common ground connection. Each circuit may have its own "ground" system, or one system may be floating.

The insulation resistance between the input and output sides is typically adequate for 1500 V or greater. Some devices are manufactured for use in analog circuits, dc to 100 kHz, while others are specially designed for pulse (digital) signal transmission. Optocouplers are available with SCRs (silicon-controlled rectifiers) on the output side to handle considerably more power than ordinary transistors. The SCR device can then be called a "solid-state relay" with optical isolation. Thus many ac loads can be directly controlled with low power signals from a computer, with no direct wire connections between.

Optocouplers are extremely useful for connecting inputs and outputs of control room electronic signal circuits to remote electronics involved with sensors and actuators. This remote equipment may have a different ac line power source, so trouble might arise with ground loops if "ground" from the remote equipment is simply connected to the "ground" at the control room. By using optocouplers there is no direct wire connection from the remote circuitry to control room equipment.

A typical application is shown in Fig. B-2; the control room receives digital signals on circuit 1 and sends out digital signals on circuit 2. The remote equipment supplies power for operating the optocouplers, so that there is no direct wire connection between remote equipment and the control room equipment.

Many control equipment vendors supply digital input and output (I/O) circuitry with optocouplers already provided in all circuits because

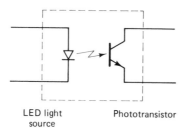

LED light
source

Phototransistor

Fig. B-1   Optical isolator.

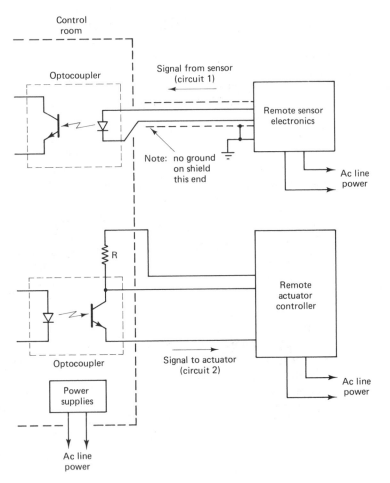

Fig. B-2   Signal isolation using optocouplers TO and FROM control room.

their industrial experience has taught them the value of circuit isolation. Other vendors are not so experienced in industrial applications and problems. It pays to ask questions and inquire about optical isolation!

The reader is referred to the Bibliography for further study of this topic.

C

Shielded Data Links and

Methods of Selection*

*Appendix C is reprinted in its entirety (including illustrations) by permission of
Electronic Products, Hearst Business Communications, Garden City, N.Y.; illustrations
courtesy of Belden Wire and Cable, Division of Cooper Industries.

"SHIELD THAT CABLE"

Bruce Morgen, Associate Editor
Electronic Products

With the FCC clamping down on EMI/RFI emissions from small computer systems, there has been renewed interest in the shielding of cables as a means of holding down system noise generation. Then there is the potentially disastrous problem of data pollution caused by ambient EMI/RFI pickup by inadequate shielding cabling. As frustrating as it is to have an otherwise good system fail to meet FCC requirements, it is equally aggravating to watch your latest design go down every time a car with a noisy ignition system goes by! While well-shielded cables are not a cure-all for such maladies, they often make the critical difference between a reliable . . . and salable . . . product and an expensive white elephant.

Shield Performance

If shield effectiveness was the only criterion for selecting shielding, you could simply run wiring through pipes with solid copper walls. Since most of us require a good deal more flexibility of our cables than that of a solid-wall copper tube, we generally settle for less-than-ideal shielding. Instead, we strive for shielding effectiveness that approaches ideal performance as closely as possible within the real-world constraints of flexibility, termination considerations, and cost. To do this, we need a practical way to measure a shield's performance at the frequency involved. While there have been many attempts to arrive at a useful measurement of shielding effectiveness, the current consensus among wire and cable experts favors transfer impedance.

Transfer impedance is the ratio of the potential difference applied at a pair of terminals in a network to the resulting current at a second pair of terminals. If the network in question is a single-conductor shielded cable, terminated with its characteristic impedance and located in an ambient electromagnetic field, an interference current (I_e) is induced in the shield (see Fig. C-1).

What Happens in the Shield?

Part of the incident electromagnetic energy is reflected from the shield and part of it penetrates the shield. This penetrating energy is subject to a degree of attenuation that is dependent on the effectiveness of the shield. Whatever energy does get through generates an interference

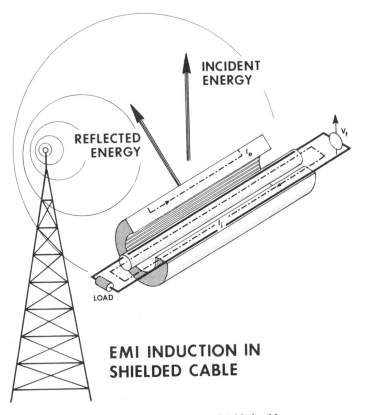

EMI INDUCTION IN SHIELDED CABLE

Fig. C-1   EMI induction in shielded cable.

voltage (Vt) on the cable's circuit and the current flowing through it (Ii). The better the shield, the lower the interference voltage for a given interference current. By calculating the ratio of interference voltage to interference current (Vt/Ie) and controlling the frequency in the electromagnetic field, we can determine the transfer impedance (Zt) of the shield at that frequency.

As is apparent from the complete equation ... $Zt = Vt/Ie$ ... the lower the transfer impedance, the better the shield. The transfer impedance is expressed in ohms per meter. The transfer impedance in ohms/m is obtained by dividing Zt by the length of the sample, in meters, or by using one meter as a standard length.

Figure C-2 shows the transfer impedance of some typical shields used for electronic cables, along with our hypothetical solid-wall copper tube and a similar tube made of stainless steel. Below 100 kHz (0.1 MHz) the transfer impedance is equal to the DC resistance. This is why the

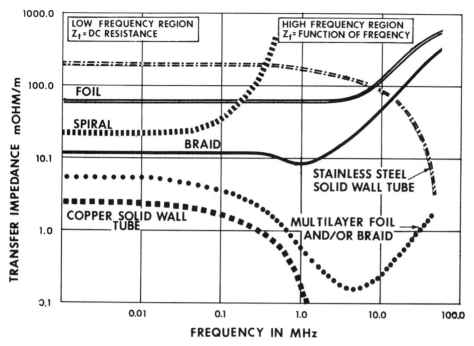

Fig. C-2 Graph of transfer impedance for various shielding methods or types.

more conductive copper tube is a better shield than a dimensionally identical stainless steel tube. The other factors influencing shield transfer impedance are thickness and magnetic permeability. As frequency increases, current tends to flow along the inside surface of the shield. This is referred to as "skin effect." Skin effect results in increased attenuation by the shield, which is observable in solid-wall shields as decreasing transfer impedance with increasing frequency.

Another common method of shield evaluation is optical coverage. This is the method used by cable manufacturers when they describe a shielded cable as having "96% coverage." Such figures are arrived at by dividing the surface area of the cable's inner conductor insulation that is covered by the shield by the total surface area of that insulation.

Shield Construction

The most flexible of the commonly available cable shields is the spiral wrapped or served shield. Spiral shields consist of bare or tinned copper strands wrapped around insulated conductor (s). Optical coverage as high as 97% is possible with this construction. The major limitation of spiral shielding is generally poor performance at frequencies above 100 kHz. This is due to the highly inductive nature of the spiral configuration, which produces longitudinal solenoid magnetic fields. These fields add to transfer impedance resulting in a curve that rises steeply with increasing frequency. Good inductors that they are, spiral shields can act as low-pass filters, generally not a desirable trait. On the plus side, spiral shielded cables are easy to terminate and quite low in cost.

Braid Is Better

Somewhat better performance is obtainable with braided shields. The shield is woven with copper (or aluminum) stranded carriers, with one set woven in a right-hand lay and the other in a left-hand lay (see Fig. C-3). Braided shield cables are not as flexible, easy to terminate, or as inexpensive as spiral-wraps. Transfer impedance at low frequencies is lower than spiral-wraps because of considerably lower DC resistance . . . the crossing strand carriers shorten the signal path. The transfer impedance does not rise significantly until well above 1 MHz because the oppositely wound strand carriers produce compensating solenoidal fields rather than a single longitudinal field generated by spiral-wrapped shields. The major limitation of braided shielding is that the weave inherently contains apertures. By carefully selecting weave parameters like wire gauge, ends per strand of carrier, picks per unit of cable length, and weave angle, optical coverage as high as 97% is possible. However,

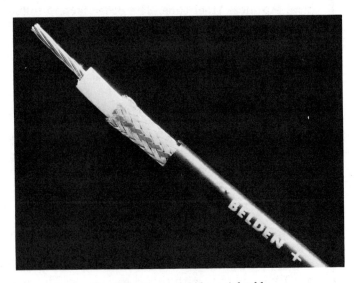

Fig. C-3   Copper braid coaxial cable.

this percentage is not sufficient to prevent transfer impedance from increasing to a measurement that can be 20 times higher at 30 MHz than it is at 10 MHz.

The most recent development in cable shielding is the foil shield (see Fig. C-4), which usually consists of aluminum foil laminated to polyester tape or adhesively bonded to the inner conductor insulation. Since the foil is only 1 to 3 microns (0.000039″ to 0.00012″) thick, the shield's DC resistance—and its transfer impedance at low frequencies— is relatively high despite the "drain wire." The foil shield is wound spirally, but the tape is wide enough to minimize inductive effects. The most promising of its virtues is that 100% optical coverage is possible. One would think that the transfer impedance of the foil shield would show a decrease with increasing frequency like the solid-wall copper tube. Instead, first generation foil shields showed a transfer impedance increase similar to that of braided shields. This is due to the lack of electrical continuity across the seams between the turns of foil shield tape. The polyester or adhesive backing, necessary for strength, was acting as an insulator and creating a spiral-shaped aperture or "slot" that allowed electromagnetic field penetration.

Since foil cables are relatively inexpensive and provide easy shield termination via the uninsulated drain wire, there was plenty of motivation for solving the slot problem. The solution arrived at was the short-

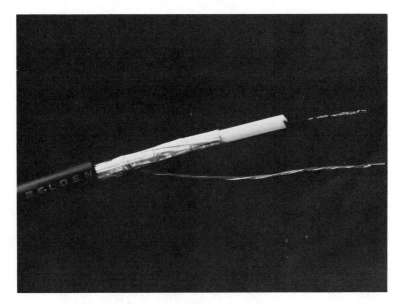

Fig. C-4   Foil shield with drain wire.

ing fold (see Fig. C-5), which effectively eliminates the slot, at least at frequencies below 10 MHz. Shorting-fold variations have been developed which allow the extra benefit of using polyester backing as an additional layer of insulation, a valuable feature if the foil-shield conductor is part of a crosstalk prone cable assembly.

Foil-shield performance can be further improved by additional drain wires and a second layer of foil on the opposite side of the polyester. Foil-shield cables with multiple drain wires and two layers of foil can have lower transfer impedance than typical braided-shield cables at 15 MHz or higher.

Belt and Suspenders

A significant trend in cable shielding combines the foil and braided approaches (see Fig. C-6). Covering a two-layer foil shield with a braided shield can result in a transfer impedance curve that falls starting at

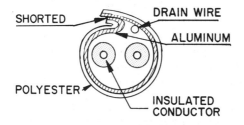

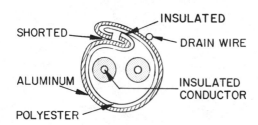

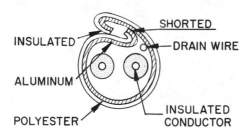

Fig. C-5 Types of foil shorting folds.

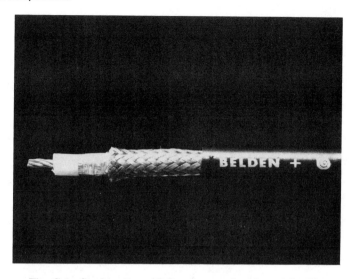

Fig. C-6 Combination of foil and copper braid coaxial cable.

between 1 and 3 MHz . . . performance that begins to resemble that of the solid-wall copper tube. This shielding approach not only out performs the best braided shields, it is also far less costly. Since the foil provides 100% optical coverage, there is no need to minimize braid apertures, and relatively inexpensive braids with optical coverage in the 40 to 80% range can be used.

If even closer approximation of soild-wall copper tube performance is desired, more layers of foil and braid can be added. A good example of this type of approach can be found in the cable used for the Ethernet local area network (see Fig. C-7). Ethernet cable has a polyester-backed foil shield bonded to the inner conductor insulation, covered by a tinned copper braided shield with 92% optical coverage, a two-foil shield sharing a common polyester backing, and a second tinned-copper braided

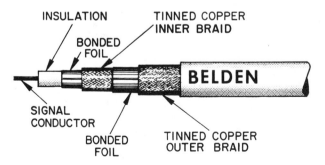

Fig. C-7 Belden Ethernet four shield coaxial cable.

shield with 92% optical coverage. Transfer impedance of this shield configuration falls steadily with increasing frequency up to 30 MHz. Three-layer shields developed for CATV coaxial cables are better yet. These designs use a bonded-foil/braid/foil-with-shorting-fold shielding format and exhibit transfer impedance curves that continue to fall until between 80 and 100 MHz.

About Emissions

If a designer's major concern is EMI/RFI emissions, rather than susceptibility, the question often arises: Does transfer impedance tell me anything about a shield's effectiveness against emissions from the cable? Since transfer impedance measurements result from the effects of an electromagnetic field originating outside the cable, they essentially measure susceptibility. Moreover, current theoretical formulae for calculating shield effectiveness against cable-sourced EMI/RFI are rather complex and generally not published by the suppliers.

Cable makers do, however, perform carefully controlled radiation tests. When viewed together with the transfer impedance curves of the same shields (see Fig. C-8), these tests show that transfer impedance measurements correlate closely enough to shield effectiveness against cable-sourced emissions to be used as a selection guideline.

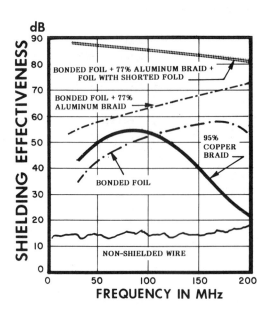

Fig. C-8 Relationship between transfer impedance and shielding effectiveness.

Shield Connection Formats

Although "grounded" cable shields, with the shield connected to circuit and chassis ground at both ends of the cable (see Fig. C-9) provide sufficient EMI/RFI protection in most applications, there are cases where this approach can lead to ground loop or common mode interference. This is caused by differences in potential between nominal ground points to which opposite ends of the shield are connected. This condition permits unwanted noise to be carried on the shield along with the signal return. This problem, which is especially troublesome at frequencies of 6 MHz or lower, can be remedied by separating the chassis and circuit/shield grounds (see Fig. C-10) and reducing the number of ground connections to the shield to a minimum. This requires that chassis-mounted cable connectors be insulated from panels and connected to the circuit ground.

It has been pointed out that neither aluminum nor copper shielding is effective against low-frequency magnetic fields. In situations where such fields are present, or where very high levels of EMI/RFI or crosstalk are a problem, cables employing electrically separated multiple shields, twisted-pair inner conductors, or a combination of the two configurations are often used.

Although no more effective than the usual coaxial shielded cable configuration against low-frequency magnetic fields, triax cable, with its two shields, provides a number of advantages. By grounding the outer

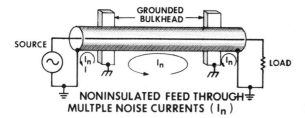

NONINSULATED FEED THROUGH
MULTPLE NOISE CURRENTS (I_n)

Fig. C-9 Shielding methods.

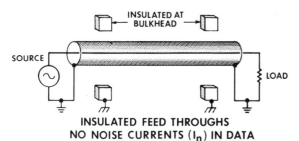

INSULATED FEED THROUGHS
NO NOISE CURRENTS (I_n) IN DATA

Fig. C-10 Shielding methods.

shield and using the inner shield exclusively as the signal return (see Fig. C-11), not only are ground loops avoided, but capacitive coupling of external noise fields to the signal carrying conductors is also prevented. Triax can also be connected in a "driven shield" configuration (see Fig. C-12). When used this way, the cables outer, grounded at the source end, is the signal return. The inner shield is connected to the center conductor at the source and acts as a Faraday shield. This has no particular advantage with respect to interference control, but it does reduce the distributed capacitance of the cable considerably. This permits longer cable runs in systems where high data rates are transmitted.

One approach that is effective against low-frequency magnetic fields is to use balanced line transmission via twisted-pair center conductors along with a shield. Twinax cable (see Fig. C-13), with its specified characteristic impedance, is an increasingly popular example of this technique. It is the least expensive cable that is effective against EMI/RFI, crosstalk, and low-frequency magnetic fields. Its major disadvantage is a frequency ceiling of 15 MHz due to signal loss.

By combining triax's two separate shields with twinax's twisted-pair center conductors, quadrax cable acquires the virtues of both. Perhaps the ultimate in EMI/RFI and crosstalk shielding effectiveness results from connecting the outer shield of the quadrax cable to the system ground and the inner shield to a true "earth" ground (see Fig. C-14).

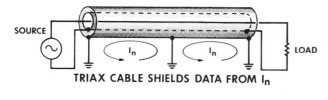

TRIAX CABLE SHIELDS DATA FROM I_n

Fig. C-11   Shielding methods.

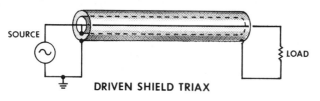

DRIVEN SHIELD TRIAX

Fig. C-12   Shielding methods.

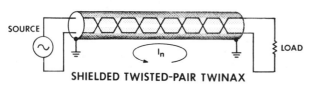

SHIELDED TWISTED-PAIR TWINAX

Fig. C-13   Shielding methods.

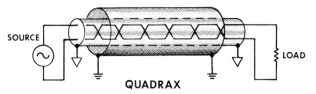

SOURCE

LOAD

QUADRAX

Fig. C-14 Shielding methods.

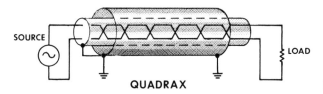

SOURCE

LOAD

QUADRAX

Fig. C-15 Shielding methods.

Where an earth ground is impractical, similar performance can be had by connecting the inner shield to the system ground at the source end only, while retaining the outer braid ground connections at both ends of the cable (see Fig. C-15). Where the low-frequency magnetic field suppression of the twisted-pair center conductors is not needed, both of these shield configurations can be used with a coaxial cable having three separate shields.

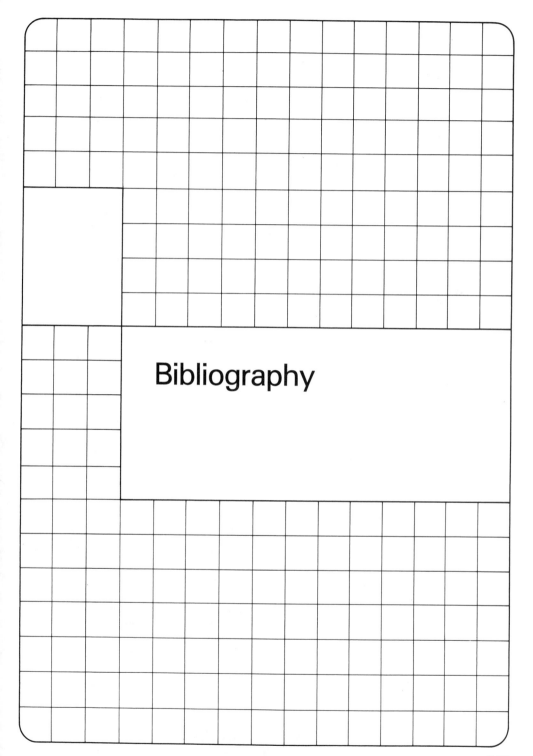

Bibliography

Actuators: Electrical, Pneumatic, and Hydraulic

PATRICK, D. and S. FARDO, *Industrial Process Control Systems.* Englewood Cliffs, N.J.: Prentice-Hall, Inc., 1985.

Analog Measurements and Interfacing

GOLDSBROUGH, P., T. LUND, and J. RAYNER, *Analog Electronics for Microcomputer Systems.* Indianapolis, Ind.: Howard W. Sams & Company, Publishers, 1983.

Closed-Loop and Feedback Control Basics

DOEBELIN, E. O., *Dynamic Analysis and Feedback Control.* New York: McGraw-Hill Book Company, 1962.

JOHNSON, C. D., *Process Control Instrumentation Technology*, 2nd ed. New York: John Wiley & Sons, Inc., 1982.

PATRICK, D., and S. FARDO, *Industrial Process Control Systems.* Englewood Cliffs, N.J.: Prentice-Hall, Inc., 1985.

WEYRICK, R. C., *Fundamentals of Automatic Control.* New York: McGraw-Hill Book Company, 1975.

WILSON, J. A. SAM, *Industrial Electronics and Control.* Chicago: Science Research Associates, Inc., 1978.

Communications

BRANT, C. A., *Electronics for Communications.* Chicago: Science Research Associates, Inc., 1983.

GURRIE, M., and P. O'CONNER, *Voice/Data Telecommunications Systems: An Introduction to Technology.* Englewood Cliffs, N.J.: Prentice-Hall, Inc., 1986.

KILLEN, H., *Telecommunications and Data Communications System Design with Troubleshooting.* Englewood Cliffs, N.J.: Prentice-Hall, Inc., 1986.

SHRADER, R. L., *Electronic Communication*, 5th ed. New York: McGraw-Hill Book Company, 1985.

Computer Basics and Industrial Applications

GINN, P. L., *An Introduction to Process Control and Digital Computers.* Houston, Tex.: Gulf Publishing Company, Book Division, 1982.

HARRISON, T. J. (Ed.), *Handbook of Industrial Control Computers.* New York: John Wiley & Sons, Inc., 1972.

Computer Control

GUPTON, J. A., JR., *Computer-Controlled Industrial Machines, Processes and Robots.* Englewood Cliffs, N.J.: Prentice-Hall, Inc., 1986.
SEAMS, W. S., *Computer Numerical Control: Concepts & Programming.* Albany, N.Y.: Delmar Publishers, Inc., 1986.

Digital Basics

GREENFIELD, J. D., *Practical Digital Design Using IC's.* New York: John Wiley & Sons, Inc., 1977.
TOCCI, R. J., *Digital Systems: Principles and Applications*, 3rd ed. Englewood Cliffs, N.J.: Prentice-Hall, Inc., 1985.

Digital Communications

ALISOUSKAS, V., and W. TOMASI, *Digital and Data Communications.* Englewood Cliffs, N.J.: Prentice-Hall, Inc., 1985.
STEIN, D. H., *Introduction to Digital Data Communications.* Albany, N.Y.: Delmar Publishers, Inc., 1985.

Electronics: Introductory

GARRISH, H. H., and W. F. DUGGER, JR., *Transistor Electronics.* South Holland, Ill.: Goodheart-Willcox Co., 1981.
HERMAN, S. L., *Electronics for Industrial Electricians.* Albany, N.Y.: Delmar Publishers, Inc., 1985.
MALCOLM, D. A., JR., *Fundamentals of Electronics.* Boston: Breton Publishers, 1983.
SHRADER, R. L., *Electrical Fundamentals for Technicians*, 2nd ed. New York: McGraw-Hill Book Company, 1977.

Electronics: Sensors and Controllers

MALONEY, D., *Industrial Solid State Electronics: Devices and Systems*, 2nd ed. Englewood Cliffs: N.J.: Prentice-Hall, Inc. 1986.
SCHULER, C. A., and W. L. McNAMEE, *Industrial Electronics & Robotics.* New York: McGraw-Hill Book Company, 1986.

Interfacing

Analog Devices, Inc., Engineering Staff, *Analog-Digital Conversion Handbook*, 3rd ed. Englewood Cliffs, N.J.: Prentice-Hall, Inc., 1986.
Analog Devices, Inc., Engineering Staff, *Synchro & Resolver Conversion.* Englewood Cliffs, N.J.: Prentice-Hall, Inc. 1980.

Analog Devices, Inc., Engineering Staff, *Transducer Interfacing Handbook*. Englewood Cliffs, N.J.: Prentice-Hall, Inc., 1980.

ANDREWS, M., *Programming Microprocessor Interfaces for Control and Instrumentation*. Englewood Cliffs, N.J.: Prentice-Hall, Inc., 1982.

ARTWICK, B. A., *Microcomputer Interfacing*. Englewood Cliffs, N.J.: Prentice-Hall, Inc., 1980.

BIBBERO, R. J., *Microprocessors in Instruments and Control*. New York: John Wiley & Sons, Inc., 1977.

CHULEY, J. C., *Interfacing to Microprocessors*. New York: McGraw-Hill Book Company, 1983.

LARSEN, D. G., J. A. TITUS, P. R. RONY, and C. A. TITUS, *Interfacing and Scientific Data Communications Experiments*. Indianapolis, Ind.: Howard W. Sams & Company, Inc., Publishers, 1979.

LUETZOW, R. H., *Interfacing Test Circuits with Single Board Computers*. Blue Ridge Summit, Pa.: TAB Books, Inc., 1983.

MALCOLM-LAWES, D. J., *Microcomputer and Laboratory Instrumentation*. New York: Plenum Press, 1984.

TITUS, J. A., C. A. TITUS, P. R. RONY, and D. G. LARSEN, *Microcomputer-Analog Converters Software and Hardware Interfacing*. Indianapolis, Ind.: Howard W. Sams & Company, Inc., Publishers, 1978.

Interference Control

MARDIGUIAN, MICHAEL, *Interference Control in Computers & Microprocessor-Based Equipment*. Gainesville, Va.: Don White Consultants, Inc., 1984.

Measurements: Alternating-Current Power

EMANUEL, P., *Motors, Generators, Transformers, and Energy*. Englewood Cliffs, N.J.: Prentice-Hall, Inc., 1985.

LISTER, E. C., *Electric Circuits & Machines*, 6th ed. New York: McGraw-Hill Book Company, 1984.

Measurements: Sensors and Transducers

ALLOCCA, J., and A. STUART, *Electronic Instrumentation*. Reston, Va.: Reston Publishing Co., Inc., 1983.

BELL, D., *Electronic Instrumentation and Measurements*. Reston, Va.: Reston Publishing Co., Inc., 1983.

HUMPHRIES, J. T., and L. P. SHEETS, *Industrial Electronics*, 2nd ed. Boston: Breton Publishers, 1983.

JOHNSON, C. D., *Process Control Instrumentation Technology*, 2nd ed. New York: John Wiley & Sons, Inc., 1982.

KANTROWITZ, P., G. KOUSOUROU, and L. ZUCKER, *Electronic Measurements*. Englewood Cliffs, N.J.: Prentice-Hall, Inc., 1979.

LENK, J., *Handbook of Practical Electronic Tests and Measurements.* Englewood Cliffs, N.J.: Prentice-Hall, Inc., 1969.

MARCUS, A., and J. LENK, *Measurements for Technicians.* Englewood Cliffs, N.J.: Prentice-Hall, Inc., 1971.

SEIPPEL, R. G., *Transducers, Sensors, and Detectors.* Reston, Va.: Reston Publishing Co., Inc., 1983.

Microprocessor Basics

BISHOP, R., *Basic Microprocessors & the 6800.* Hasbrouck Heights, N.J.: Hayden Book Company, Inc., 1979.

BRUNER, H., *Introduction to Microprocessors.* Reston, Va.: Reston Publishing Co., Inc., 1982.

BURTON, D. P., and A. L. DEXTER, *Microprocessor System Handbook.* Norwood, Mass.: Analog Devices, Inc., 1977.

GREENFIELD, J. D., and W. C. WRAY, *Using Microprocessors and Microcomputers: The 6800 Family.* New York: John Wiley & Sons, Inc., 1981.

TOCCI, R., and L. LASKOWSKI, *Microprocessors and Microcomputers: The 6800 Family.* Englewood Cliffs, N.J.: Prentice-Hall, Inc., 1986.

Microprocessor Applications

JOHNSON, C. D., *Microprocessor-Based Process Control.* Englewood Cliffs, N.J.: Prentice-Hall, Inc., 1984.

STOUT, D. F., *Microprocessor Applications Handbook.* New York: McGraw-Hill Book Company, 1982.

Motors, Transformers, and Controls

ALERICH, W. N., *Electric Motor Control*, 3rd ed. Albany, N.Y.: Delmar Publishers, Inc., 1983.

EMANUEL, P., *Motors, Generators, Transformers, and Energy.* Englewood Cliffs, N.J.: Prentice-Hall, Inc., 1985.

Nuclear Power

LYERLY, R. L., and W. MITCHELL, III, *Nuclear Power Plants.* Washington, D.C.: USAEC, rev. 1964.

U.S. Department of Energy, *Nuclear Power from Fission Reactors.* Washington, D.C.: U.S. Department of Energy, 1982.

Operational Amplifiers

DUNGAN, F. R., *Linear Integrated Circuits for Technicians.* Boston: Breton Publishers, 1984.

FAULKENBERRY, L. M., *An Introduction to Operational Amplifiers with Linear IC Applications*, 2nd ed. New York: John Wiley & Sons, Inc., 1982.

HUMPHRIES, J. T., and L. P. SHEETS, *Industrial Electronics*, 2nd ed. Boston: Breton Publishers, 1983.

Programmable Controllers

JONES, C. T., and L. A. BRYAN, *Programmable Controllers.* Atlanta, Ga.: International Programmable Controls, Inc., 1983.

KISSELL, THOMAS E., *Understanding and Using Programmable Controllers.* Englewood Cliffs, N.J.: Prentice-Hall, Inc., 1986.

Steam Boilers

The Babcock & Wilcox Co., Inc., *Steam: Its Generation & Use*, 37th & later eds. New York: The Babcock & Wilcox Co., Inc., 1963.

Troubleshooting

LENK, J., *Handbook of Practical Microcomputer Troubleshooting.* Reston, Va.: Reston Publishing Co., Inc., 1979.

LOVEDAY, G., and A. H. SEIDMAN, *Troubleshooting Solid State Circuits.* New York: John Wiley & Sons, Inc., 1977.

OLEKSY, J. E., *Electronic Trouble Shooting.* Indianapolis, Ind.: The Bobbs-Merrill Company, Inc., 1983.

PEROZZO, J., *Microcomputer Trouble-shooting.* Albany, N.Y.: Delmar Publishers, Inc., 1986.

PEROZZO, J., *Practical Electronic Trouble-Shooting.* Albany, N.Y.: Delmar Publishers, Inc., 1985.

ROBINSON, J. B., *Modern Digital Troubleshooting.* Redmond, WA: DATA I/O Corp., 1983.

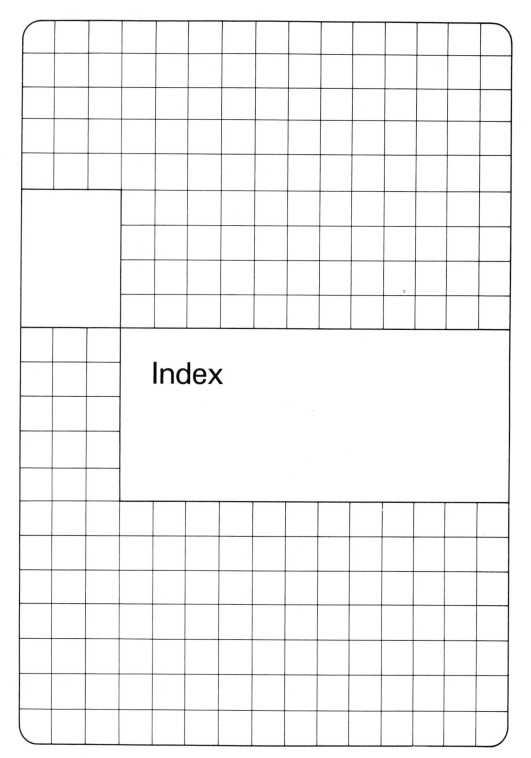

Index